MOULDMAKING AND CASTING

Nick Brooks

THE CROWOOD PRESS

First published in 2005 by
The Crowood Press Ltd
Ramsbury, Marlborough
Wiltshire SN8 2HR

enquiries@crowood.com

www.crowood.com

Paperback edition 2024

British Library Cataloguing-in-Publication Data
A catalogue record for this book is available from the British Library.

ISBN 978 0 7198 4401 0

Disclaimer
Safety is of the utmost importance in every aspect of mouldmaking and casting work. When using tools and/or chemicals, always closely follow the manufacturer's recommended procedures. The author and publisher cannot accept responsibility for any accident or injury caused by following the advice given in this book.

Typeset by D&N Publishing, Baydon, Wiltshire

Cover design by Maggie Mellett

Printed and bound in India by Thomson Press

Dedication
To Clementine and Talia, my best work ever.

Acknowledgements
Clementine, Talia and Dan for technical assistance.
Clive and Diana.
Jack Ludlum (www.londonpaparazzi.com) for photographic contributions.
Simon and Anna Cooley for sculpting and modelling clay portrait head in Chapter 4.
Tony Lall-Chopra and Sabine Heitz for mouldmaking and modelling in Chapter 9.
Central Saint Martins' staff for encouragement and support.
Norman Balon for C&H services.
Brendan and Vicki.
Buzzcocks for broken ankle.
Praveera for being there at the beginning.
Alec Tiranti Limited – Sculptors' tools, materials and equipment, established 1895, for supply of all mouldmaking and casting materials used in producing this book.

About the author
Nick Brooks studied Fine Art and Design at Middlesex University, then travelled and worked abroad extensively in the 1980s. He has experience in all types of mouldmaking and casting, having worked in the British film industry and accepted a range of private and commercial commissions. He has been teaching mouldmaking, casting and sculpture for over twenty years institutions around the UK including, Central Saint Martins College (Central London), West Dean College (West Sussex), Study Sculpture (Brecon Beacons), Broadlands Art Centre (East Anglia) and Yorkshire Coast College (Scarborough).

Also by Nick Brooks:

CONTENTS

'Leaf'. By Antigoni Pasidi. Silicone rubber, life-size.

PREFACE

I have taught and practised mouldmaking and casting techniques and processes for many years, and have been encouraged to write something on the subject many times. It is not an easy subject to put into the written word – the multiple possibilities and unique requirements of any individual project potentially present as many multiple alternative possibilities. However, there are a number of fundamental techniques that can be taught; these represent a starting point for anyone interested in the subject, and can then be adapted, combined and even potentially 'reinvented' to an individual project (there are few 'rules'!).

When considering this project, I came to the conclusion that the main requirement was for a book that would explain and illustrate the principal techniques and processes. These techniques and processes may be covered elsewhere individually, but are seldom to be found combined in the one book. I have learnt many of the techniques and principles contained here formally but then adapted them over the years. In my opinion, this is the purpose of any acquisition of knowledge, to make it one's own.

Whether you use this book as an introduction to a new subject or to further existing knowledge, I hope you will find the creative scope that mouldmaking and casting offers as exciting and fulfilling as I have and do.

Nick Brooks

CHAPTER 1

HISTORY

The Beginning of Mouldmaking

Long before humans walked the Earth, countless species of animals stamped their footprints in the mud. Fossilization has turned some of these footprints into moulds from which present-day anthropologists, using modern plastics, are able to make casts for scientific records and museum exhibits. The Earth itself was therefore the first mould.

To recreate the visual world in three dimensions was an astonishing cognitive leap for humanity. Many creation myths name clay as the material from which gods or goddesses made the first humans and animals. We know that Upper Palaeolithic hunter-gatherers made small, portable human and animal figurines, usually of ivory, stone, bone or clay. One of the earliest known baked clay sculptures, the 26,000-year-old Venus figurine found at Dolni Vestonice in Moravia, shows that clay was fired at an early date.

The ritual burial of objects of cult significance, often in the form of small animal and human figurines, has occurred in almost every culture throughout the world. The Ancient Egyptian Ushabti figures, the remarkable figures from the island necropolis of Jaiana of the classical Mayan period (AD700–800) and the more recent painted terracotta Mexican sculpture are some examples. The urge to make the 'one' into the 'many' has long inspired humanity to replicate those objects found to be beautiful or useful. A proficiency in sculptural mouldmaking generally appeared in societies where organized religion or burial customs required votive images, symbolic objects as offerings to the gods and tomb figures to be buried with the dead. Initially these were handmade by the

women or shaman of the tribe, but with the coming of more formalized religious practices the demand for repeated designs increased and as a result various techniques of moulding were used to speed up the output. In many cases, moulding was used in conjunction with hand modelling of faces, hands and other details so as to individualize each figure. The craft became specialized, often resulting in a high degree of artistic production.

An example of such a sequence of events, spanning many centuries, can be seen in China, where an impressive Neolithic ceramic tradition paved the way for the unparalleled skill in moulding of the Shang Dynasty (1523–1027BC). Human sacrifice originally played a prominent part in burial traditions, but when the practice was outlawed tomb figures were required in large numbers as symbolic substitutes. The most famous example in China is the terracotta army of Shih Huang Ti at Xian, c.200BC, in which the trunks are moulded but no two faces are the same. The beautifully expressive terracotta Tang figures of around the eighth century AD were perhaps the ultimate artistic achievement in the production of tomb figures. For these, both press moulds and multiple moulds were used, with the figures and animals being finished by hand modelling and the use of richly coloured glazes. The same techniques were used for other sculpture intended to provide inspiration for the living. The larger-than-lifesize Lohan Buddhas of the tenth to thirteenth centuries AD were possibly moulded from an admired original, then individualized by modelling before being fired, often in one piece – an impressive feat for such a large statue.

The discovery of new materials ensured the continued development of ceramic techniques. Around the fourth century AD the Chinese invented stoneware, and by the eighth century porcelain had been perfected. A wide variety of techniques again combined moulding with modelling to produce porcelain

OPPOSITE: 'Augustus'. Ancient Roman emperor. By Nick Brooks. Cold-cast resin bronze. H 27cm, W 17cm, D 10cm.

statuettes for domestic shrines. These figures in white porcelain, which came to be known in Europe as *blanc de chine*, were among the most prized works of Eastern art in Europe in the early eighteenth century. Copies of small figures of the Buddhist Bodhisattva Gwanyin were amongst the earliest works sold by the Royal Saxon Porcelain Manufactory at Meissen.

The ceramic traditions of China, though rarely surpassed, had impressive parallels in ancient Mediterranean civilizations. In seventh century BC Greece the growing importance of sanctuaries such as Delphi and Olympia created an increasing demand for votive offerings. When the small figurines originally made for votive offerings became a popular secular genre it was necessary to speed up production. At first, just details or parts were stamped or press moulded; next, use of a two-part mould enabled the front and back to be moulded separately and joined with slip to form a rounded image. Finally, the use of multi-part press moulds meant that the parts could be assembled when the clay was leather hard in any variety of poses and gestures. The *koroplasts* who made the patterns were not only fine craftsmen but truly creative artists as well who could demand high prices for their work. Both finished figures and moulds were exported widely around the Mediterranean area.

Press Moulding

The more practical uses of casting and moulding also date back many thousands of years. Stamp seals carved in intaglio and impressed on clay have been dated from *c*.7000BC. These were the origin of the use of a positive and a negative. By the late fourth millennium BC Sumerians were using cylinder seals, and the seal cutter, or *purkullu*, was a high-ranking professional craftsman who had to complete a four-year apprenticeship. The subjects depicted on seals have proved a mine of information for historians. Seals were put to many uses. Ownership could be established, property could be marked. Clay-stamped ropes were used to seal the doors of storage areas. Seals were also amulets and had a magical as well as a practical purpose. They were used in rituals for symbolic sealing. Figures of pregnant women were 'sealed' to ensure the safety of the unborn child, and clay tongues or lips were 'sealed' to indicate the need for secrecy. The legal and administrative use of the seal extends to this day.

The use of clay to make pottery vessels appears at widely varying times in different parts of the world, but generally coincides with the abandonment of the Palaeolithic hunter-gatherer culture for a more settled agricultural lifestyle. Neolithic people caulked their baskets with clay to store grain, and so the burning of discarded baskets may have led to the development of the moulding and firing of pottery vessels. Large shells and skulls served as moulds for smaller clay vessels. Pottery fragments with pressed cord decorations at later settlements in Jarma and Jericho (7000BC) suggest an early link between clay with the lighter portable basket. Handmade coil and slab pots can be found in every early ceramic culture, but the concept of using a mould to achieve standardization and high productivity, though obvious to us, was for early humanity a great technological step forward.

Although the potter's wheel was known by 4000BC in Mesopotamia, its use spread slowly and a variety of other ceramic techniques continued to be used in other places. The Ancient Egyptians pressed clay over a core of wood to produce plates and bowls in the Early Dynastic period (3000–2600BC). In the New World, where the wheel was a later development, Pre-Columbian potters developed a high degree of skill in the use of moulds. The Romans, who brought the wheel to Britain, also brought their knowledge of forming bricks in moulds. Although more complex multipart moulds and other techniques were developed, the simple technique of press moulding continued to have its uses throughout the world for many centuries. It continued to be used for architectural details until the nineteenth century AD, for example on the decorated terracotta façades of European buildings.

Metal Casting

The way in which expertise in working one material can further the development of a technique in another can be observed in the influence that metal casting had on mould-making. Although the natural progression within the development of metal working, from hammering, annealing, smelting to casting, took the same course in the early metal technology of various cultures, it followed different time scales. Sometimes a considerable time elapsed before a connection was made between concept and practical application. Carved stone seals, which harness the idea of positive and negative, were in use at Cayonu in Turkey (*c*.7000BC), at about the same date as objects hammered from native copper, and yet there is no evidence of metal casting in the Middle East until more than 4,000 years later. The oldest known metal casting is a copper frog made in Mesopotamia around 3200BC. By this time copper was more plentiful, as the ore was being mined, smelted and exported both in the Near East and in the Balkans.

The appearance of bronze around 3000BC led to further development of casting techniques. As the alloy was initially scarce, costly and much in demand for weapons and tools, solid-cast sculpture started mostly on a small scale. By the time of the New Kingdom in Egypt, larger statues were being cast in bronze around a clay core. During the XXII–XXIV Dynasties greater skill succeeded in achieving thinner casts. By the eighth century BC heavily leaded bronze with a low melting point was being used to mass-produce votive statuettes that were often inlaid with gold or silver. According to tradition, the first hollow-cast statues in bronze were made by Rhoecus and Theodorus of Samos in the seventh century BC, who were said to have learned their craft in Egypt. Two life-sized bronze statues of warriors found in the Ionian Sea in 1970 were made around the fourth century BC, and are amongst the finest examples ever of bronze casting.

An earlier tradition of metallurgy existed in the East. It is thought that bronze was introduced into China from the West around 2000BC. The Shang Dynasty craftsmen soon discovered that the core casting method was more economical than solid casting, and were assisted in this by their ability to make elaborate moulds in clay. Piece moulds were taken from prototypes of carved wood or pottery and large numbers of moulds were produced for the foundries. This was an early example of the Chinese use of the production line. But though mass-produced, the Shang bronze ritual vessels are undoubted masterpieces. They demonstrate a skill in casting, which, although possibly equalled by the Greeks, has never been surpassed. The expertise of Shang craftsmen in the making of reusable piece moulds may account for the fact that no evidence has been found of the use at that date of waste mould methods of casting such as the cire perdue (lost wax) method. This method was certainly known by the Sumerians, by the Indus Valley Civilization in the third millennium BC, and by the Egyptians from the XVIII Dynasty. The bronze statuette of King Tuthmosis IV (c.1385BC) in the British Museum is hollow cast around a clay core. Cire perdue was used extensively by jewellers and goldsmiths for all intricate, finely detailed work.

In the New World where the goldsmith's art reached a high degree of perfection, there are accounts by Spanish colonists of the casting of gold by the lost wax process. The Renaissance Italians also excelled in this technique, often casting very large statuary in separate sections on the ground before assembling them. The best description of this process is to be found in Benvenuto Cellini's sixteenth-century *Autobiography*, in which he described the casting of his masterpiece statue of Perseus.

Although the lost wax process remained the favoured method for the casting of bronze sculpture, many other methods have also been used. Sand casting, a simpler and cheaper process, was used in China as early as 645BC; cast-iron plough shears dating from 233BC have been found. Although only relatively simple forms could be cast in sand, the method continued to be used for weapons, domestic vessels and cooking pots, and in more recent centuries for decorative cast ironwork. The process was refined and developed by the skilled use of tooling and burnishing during the eighteenth and nineteenth centuries AD, particularly in France where bronze and brass were used for decorative mounts on vases, and on handles and feet on furniture. In Germany it was developed for editions of bronze and iron sculpture. Today it is used mainly industrially for the casting of machine parts.

The casting of sculpture in metal was revolutionized by the invention in c.1860 of electroforming. This process uses an electric current to transfer particles of metal from a bar suspended in an electrolytic solution, so as to build up a deposit on the face of a mould until sufficient thickness has been established. This technique has made possible the creation of very fine and accurate casts.

Plaster Casting

Plaster was an ideal material for making moulds and casts. In its liquid state it could be poured into moulds of wood, clay or plaster. From the nineteenth century flexible moulds of gelatin were used, and from the twentieth rubber and plastics. Plaster has been used for casting from life or for death masks, for architectural decorations, and as a mould material for casting bronze or porcelain. There is evidence of plaster being used as a building material over 9,000 years ago. Neolithic houses at Abu Hureyna in Syria have plastered floors. At shrines in Neolithic houses at Catal Huyak in Anatolia walls are decorated with plaster reliefs, and at Jericho, buried skulls have been found with faces modelled in plaster. Gypsum painted on the banks of the henges at the Neolithic site of Thornborough in Yorkshire is thought to have been used to reflect the sunlight or moonlight during rituals. The Ancient Egyptians used plaster on pyramids, and were probably the first to use it for mouldmaking. The earliest death mask is believed to be that of King Teti (c.2400BC), and plaster casts found in the studio of Tuthmose, the chief sculptor to Akenhaten, at Amama (c.1340), are presumed to have served as sculptor's models. The use of plaster casts as an aid to sculptors was further developed in the fourth century BC by Greek sculptors, who, stimulated by the popularity of the new art of portraiture, perfected the technique of piece moulding.

The Greeks were probably the first to use the process of 'pointing' for mechanically transferring the proportions from a plaster cast of a model made in wax or clay onto a block of stone or marble before carving. Pliny describes this change of technique from a direct to an indirect method of stone carving in Natural History (xxxv.153). He writes that the brother of the celebrated sculptor Lysippus, Lysistratus of Sicyon, was the first to make plaster casts from the living subject and he instituted the practice of making final corrections on a wax model taken from the plaster. The same man also invented taking casts from statues, and this practice increased to such proportions that no figure or statues were made without 'the clay' (a clay model). The Romans, who held Greek sculpture in high regard, cast many thousands of Greek sculptures. Archaeological excavation during the 1950s in the area around Naples, at Baiae, revealed the remains of a first-century AD workshop belonging to a manufacturer of plaster casts of Greek statues for wealthy Roman patrons. A well-known example of this, the Apollo Belvedere, is a Roman marble copy of a fourth century BC Greek bronze. In fifteenth-century Florence the coexistence of a renewed interest in portraiture with a demand for replicas of ancient Roman statues led to the development of the technology of bronze casting and a revival of the use of plaster casts and piece moulding. According to Giorgio Vasari (1511–74), Andrea Verrocchio (1435–88), who was much involved in bronze casting, also pioneered the making of plaster casts from life and promoted the fashion for death masks.

For many hundreds of years plaster and stucco were used as a building material and for decorative architectural mouldings, but the production of ceramics remained a workshop-based craft. Wheel-thrown pots developed their own national character through styles of decoration and glazes rather than techniques of manufacture. By the end of the seventeenth century increasing trade and travel intensified the demand, particularly amongst the aristocracy, for novelty or foreign goods and influenced local styles in sculpture, ceramics and architectural decoration. The need to cater for this demand led to two discoveries that were to be of the utmost importance in the field of ceramics in the eighteenth century. The secret of making porcelain, so long the exclusive knowledge of the Chinese, was discovered by Johann Friedrich Bottger at Meissen in 1708. In 1740 the introduction of plaster moulds by Ralph Daniel into Staffordshire meant that the mould no longer had to be made laboriously from clay, then fired. The porosity of plaster meant that even liquid clay would lose enough water to form firm walls inside the mould. With plaster moulds, slip casting therefore became possible.

As a result of these developments, ceramic production began on an industrial scale and in Britain it was to play a major role in the Industrial Revolution. Factories making soft-paste porcelain were set up in many areas and the innovations by people like Josiah Wedgwood, who opened his first factory in 1760, were to revolutionize the production and marketing of china in Staffordshire. (Ironically, the thriving international market created at that time is now threatened with extinction by competition from the East.) The finely modelled and decorated porcelain figures first produced in Meissen were soon imitated by almost every European factory of note. The moulds were made from an original master model made of wax, clay or wood. When undercutting was necessary, the figures were usually cast in separately moulded sections. The work of master pattern-makers such as Johann Joachim Kändler (1706–75), the most notable of the artisans engaged in this work at Meissen, imparted a new artistic dimension to ceramic art, just as the Greek koroplast had done centuries earlier.

Replication Through Casting

One of the great virtues of casting works of sculpture is the ability to create an identical replica of an admired but unobtainable original. The perfection that could be achieved by the use of plaster began to make it an acceptable material amongst collectors. Louis XIV had set a fashion for casts of antique sculpture in the late seventeenth century that had spread to England by the eighteenth century. The aristocracy found it to be a satisfactory alternative to expensive copies in marble by sculptors such as Canova and Thorwaldson. Casts in plaster were now valued in their own right and by the early nineteenth century casts were being taken in situ of Classical sculpture and architectural decorations and brought back in great numbers by connoisseurs and collectors. In 1816 the purchase of the Elgin Marbles generated a series of requests for casts from European museums and although these were not always granted, casts of the Marbles and other such antiquities were sometimes used to engender good public relations or to resolve the occasional restitution issue. In 1848 the British Museum paid £400 for casts of the Elgin Marbles to be sent as a gift to the King of Greece. In 1870 the 'International Convention for Promoting Universally Reproductions of Works of Art for the Benefit of All Countries' gave the final stamp of approval to the making of facsimiles in plaster.

The British Museum 1856 catalogue lists a huge variety of casts for sale. A Caryatid from the Erectheum cost £6, whereas a panel of the Parthenon Frieze could be bought for only £1 and busts generally cost from between 10/- to £1. The Natural History section included an Asiatic Elephant's tooth for 14/- and the foot of a Dodo, a bargain at 2/-! The Museum's casting department seemed to have been admirably

Second-generation cast section of the Parthenon Frieze. Greece, fifth century BC.
By Nick Brooks. H 103cm, W 125cm, D 10cm.

non-commercial and more concerned with such issues as their moral duty to keep prices affordable to educational institutions and to maintain a very high quality of casting in spite of a huge demand from all over the world. They were also very aware of the irreversible damage that could be caused to originals by the taking of casts. It was noted that any remaining traces of colour may have been lost from the Elgin Marbles when 'the whole surface of the Marbles had been twice washed over with soap lyes as that, or some other strong acid, is necessary for the purpose of removing the soap which is originally put on the surface in order to detach the plaster of the mould.' The department insisted on a faithful copy of the original and any attempt at reconstruction, however scholarly, was out of the question.

By 1840 the British Museum had direct control over the casting of its antiquities, but it was not until 1881 that it employed a permanent member of staff to make casts. Previously, the work had been done by firms of mouldmakers on a freelance basis and in order to fulfil the increasing demand this still continued. One such firm, Brucianni's, which set up in 1837 in Covent Garden, became the Museum's chief supplier of casts and by 1867 its catalogue boasted 'copies of all the finest statuary in existence' and offered a huge selection of 1,200 items. It was probably Brucianni's that made the famous 48cm fig leaf for the cast of Michelangelo's David to protect the modesty of visiting female dignitaries to the Victoria & Albert Museum. This huge 18ft cast, an unwanted gift from the Grand Duke of Tuscany, was presented by Queen Victoria to the Museum in

'House'. Sculpture of inverted house by Rachel Whiteread by permission of the artist and Gargosian gallery.

1857, and was displayed in the Architectural Courts that were opened to the public in 1873. Casts of antiquities were also in great demand in the new art schools where drawing from the 'plaister' occupied a large part of the students' time. The Royal Academy Schools were established in 1769 and during their early years gifts and acquisitions came from the collections of many leading artists and English aristocratic patrons of the arts. A survey of the cast collection of the RA Schools undertaken by John Flaxman in 1810 lists 250 'principal pieces'. These were joined in 1816 by a generous gift from the Prince Regent of 100 casts from his collection. Until the 1960s drawing from plaster casts of antique sculpture was an important part of the curriculum in European art schools, some of which still have collections of fine casts often forgotten and collecting dust in their basements.

The great virtue of casting is the creation of an identical replica. Equally, the danger of this facility has often been the multiplication of florid and tasteless forms and an inevitable loss of direct artistic expression. The feeling, often justified, that a reproduction was an intrinsically inferior production lacking in artistic expression began to have its effect on ceramics in the latter part of the nineteenth century. Influential reformers such as John Ruskin and William Morris and the Arts and Crafts Movement protested against the increased mechanization of the production of decorative objects and mourned the demise of the artist/craftsman. However, it soon became apparent that Morris's romantic insistence on the ideals of fine craftsmanship and well-designed objects for the masses was an impractical dream, and the artist/craftsman could only cater for a few wealthy patrons. In Europe the design of decorative objects tended to culminate in stylistic cul-de-sacs such as Art Nouveau. Finally, it was left to the industrial designer to come to terms with the machine in order to produce well-designed articles for the masses.

A similar change was taking place in museums. In response to the growing enthusiasm for casts, the V&A opened its new cast courts to the public in 1873. But the increasing perfection of casting techniques brought with it a need for a clear distinction to be made between reproduction and original. People wanted to be sure they knew what they were looking at. The practice of using casts to fill the gaps in a display sequence of, for instance Greek sculpture, was no longer found acceptable.

By the end of the nineteenth century sculptors and potters had begun to learn and profit from industry. By adopting the new techniques and materials used in industrial processes for moulds and casts they were able to produce increasingly perfect results. Flexible moulds first made of gelatin and later rubber compound or synthetic materials could be pulled away easily from undercut and complex surfaces without damage to the original model or cast. It was not until the 1950s that gelatin moulds were superseded by Vynagel/Vinamould (a vinyl moulding compound). By 1961 even the admirably cautious British Museum cast service was using it. The minutes of that year record that 'It has outstanding advantages over gelatin of enabling an unlimited no. of casts to be obtained from the same moulds, in contrast with gel moulds which can only provide three or four casts', and in 1962 it was noted that 'The new material described in last year's report has proved to be very durable and new moulds are being made of this wherever suitable.' Silicone rubber was, it seems, introduced shortly afterwards. Minutes in 1965 record that silicone rubber moulds 'are being made in the laboratory with Dr Werner's kind permission ... Dr W thought it possible that under his supervision Mr Prescott could teach the making of silicone rubber moulds to Mr Langhorn of the casting shop within a week.' This was obviously a success as in 1966 a memo refers to 'combining the cast service working in plaster with that working in resins'. Many artists of the latter part of the twentieth century and the early twenty-first have used aspects of mouldmaking and casting in their work, continuing many of the ancient techniques developed over the centuries and adding some new ones.

Synthetic resins and fibreglass which had been used in industry for car and boat bodies have, since the 1950s, developed into an important material for sculpture. Light, durable and relatively easy to cast they can be painted or used in conjunction with powdered metal fillers in order to produce cheap 'cold-cast' substitutes for bronze and aluminium. Fibreglass can also be used for making moulds and to provide support for flexible moulds of rubber or vinyl. In the mid 1960s Duane Hanson used polyester resin and fibreglass for his life-sized super-real sculptures which were cast from human models.

Rachel Whiteread's conceptual use of casting and moulding brings us full circle. When in 1993 she made her Turner Prize-winning work *House,* the piece drew crowds and made newspaper headlines throughout its two-and-a-half-month life. She used a Victorian terrace house as a waste mould for a concrete cast of the inside empty space.

Laser copying of computer-generated patterns is the latest addition to the mould medium and may one day threaten to replace the more traditional methods of reproducing the art object, but I somehow feel it will become yet another addition to the ancient desire to reproduce and conserve the three-dimensional creative expression.

'Section'. By Nick Brooks. Plaster.

CHAPTER 2

MOULDMAKING AND CASTING PRINCIPLES

The reasons for making a mould to produce a casting may be many, as can be seen in the ancient history of the techniques and their evolution. The artist may want to reproduce a pattern in a more durable material, for instance a clay pattern made into a plaster one. It may be desirable or necessary to produce multiples of the pattern, either for commercial possibilities or to preserve a record of the pattern. It may be easier, cheaper and more accurate to produce a mould and cast of a pattern than to try to copy it by hand modelling, for instance architectural or mechanical reproductions. All of these requirements can be satisfied with the production of a mould and cast, of which there are many types. The choice of which to use will largely depend on what the requirements are.

The following form the essence of mouldmaking:

- the mould – a shaped cavity used to give a definitive form to fluid or plastic material;
- the cast – an object shaped/formed in a mould;
- the pattern – a model used to make a mould (the word 'pattern' will be used throughout this book to describe the original form to be reproduced through a mould and cast. The pattern may be a pre-made form, for instance an industrially made object, or handmade by the artist, for instance a form made in clay).

Initial Considerations

Before embarking on any mouldmaking and casting project it is important first to consider the requirements, principles and materials of the pattern, mould and cast you will be working with. An understanding of the basic principles will help you to decide which mould and cast are best for the job.

- **Pattern material** This is the material from which the original pattern is made. It will determine what the moulding options are, and the requirements of its production.
- **Pattern complexity** The complexity and detail of the pattern will determine which type of mould will be easiest and most accurate to produce and use to cast from.
- **Casting material** This is the material in which the pattern is to be reproduced. It will have a bearing on which type of mould will be most suitable.
- **Cast detail** The level of reproductive detail reproduced from the pattern will be determined by the choice of mould being used. What degree of detail is required? For instance, is the cast to be reworked once it has been taken out of the mould? If so, a fairly crude moulding process may suffice.
- **Cast run** This is the number of castings required from the mould. Some moulds are reusable, and can be used to produce multiple casts, whereas others are 'one-offs'. Different moulds will produce different numbers of castings before detail starts to be lost, although this is also dependent upon what casting material is being used for the mould.
- **Release agent** This is a medium that is applied to a pattern to prevent the moulding material sticking to it. A release agent is also applied to a mould surface to prevent a casting sticking to it.

Basic Principles of Mouldmaking

An understanding of the essential principles of moulding is vital if the outcome is to be successful. These are outlined below.

Positive to Negative to Positive

This sequence is the fundamental working principle of mouldmaking and casting. The original pattern (positive form) is moulded (negative form of the original pattern) and then cast (reproducing the original positive form) (*see* diagram 1).

Undercutting

These are the points on a pattern where it would become 'locked' into a mould. This can best be described using a sphere as an example. If a mould were to be taken from a spherical pattern the point at which it would 'undercut' would

be at any point past the halfway line. Unless a mould were made in more than one piece, the pattern could not be removed, and therefore would be unusable to produce a cast. Furthermore, unless the line between the pieces of a two-piece mould were made exactly to the halfway line of the pattern, one or other of the mould pieces would be undercut. The way around this is either to produce a mould in more pieces, working around the undercutting, or to use a flexible mould that will 'bend' around the undercutting. A pattern with no undercutting could theoretically be moulded in a one-piece mould; however, in reality it would sometimes still be necessary to make the mould in more than one piece, to ensure removal of the pattern and subsequent casts (*see* diagram 2).

Division Lines

These are the lines on the pattern to which the mould pieces will need to be made, in order to be able to remove them past any undercutting. Once the points of undercutting have been determined, these lines should be carefully worked out and

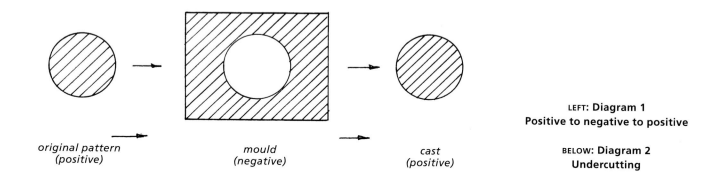

original pattern
(positive)

mould
(negative)

cast
(positive)

LEFT: **Diagram 1**
Positive to negative to positive

BELOW: **Diagram 2**
Undercutting

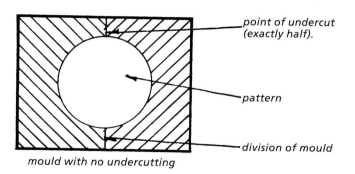

point of undercut
(exactly half).

pattern

division of mould

mould with no undercutting

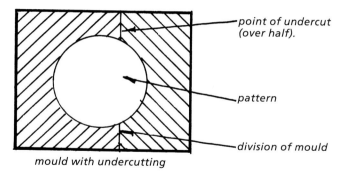

point of undercut
(over half).

pattern

division of mould

mould with undercutting

either committed to memory or marked temporarily on the pattern. The division lines do not need to be straight, but can undulate following the undercutting of the pattern (*see* diagram 3 a, b, c).

Clay Bed

A temporary bed of clay is used to divide and mask a pattern into the separate sections it will be moulded in. Once the division lines on a pattern have been established the pieces of the mould can be made one at a time. While making each mould piece, the others will need to be masked off, and this can be done by building a clay bed up to the division lines of the piece being made. As each piece of the mould is completed the clay bed is removed and built up to the lines of the next piece to be made (*see* diagram 4).

Diagram 4
Clay bed

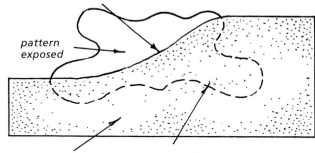

the line of division may undulate depending on pattern undercutting

pattern exposed

clay bed   pattern masked by clay bed

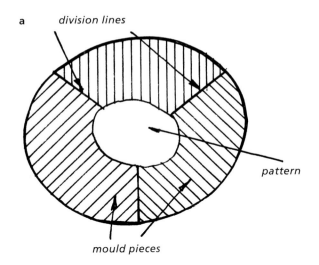

a   *division lines*

pattern

mould pieces

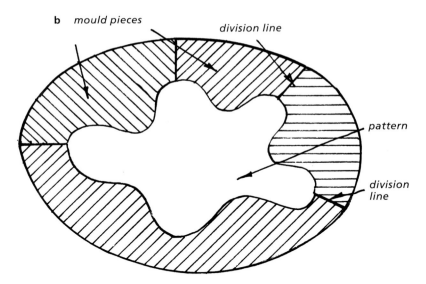

b   *mould pieces*

division line

pattern

division line

examples of a pattern and division lines required for multi-piece moulds to avoid undercutting

c   *division line showing undulating line*

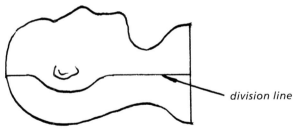

division line

Diagram 3
Division lines

17

Access into the Mould

Once the mould has been made there will need to be some sort of access into it, in order to fill it with the casting material. If the pattern has a 'footprint' (a base or area from which it stands), this will be the obvious place to become the opening into the mould. The mould is built up around this area, leaving it uncovered by the moulding material. When the mould is inverted this area will then become the 'mouth' of the mould. The size of the footprint will determine how the mould can be filled (*see* diagram 5).

If the mould is in pieces and the footprint is large enough it can be filled with the mould pieces assembled, via the mouth of the mould; otherwise, the mould will have to filled with the pieces separated and joined afterwards. In either case, the mouth of the mould needs to be at least large enough to pour some of the casting material into the mould either to fill it or to join separately filled pieces of the cast. If a pattern has no footprint and the mould is to be made completely 'in the round', a pour point will need to be built into the mould. This can be done by building a stalk of clay onto the pattern, around which the mould can be built. When the mould is completed, the stalk of clay can be removed, leaving a hole through which the casting material can be poured (*see* diagram 6). Clay stalks should be positioned on an area of the pattern where the loss of detail at that point will be least visible or easily remodelled. Stalks of other materials can be used as long as they can be temporarily securely attached and are treated with an appropriate release agent in order to be able to remove them from the moulding material when set.

Registration

Registration between mould pieces is important in order to be able to relocate accurately the pieces when casting. Initially, the shape of the mould pieces should be considered when creating masking beds or walls. Beds and walls should be created perpendicular to the pattern surface to avoid 'feathered' edges that will create weak inner edges between mould pieces (*see* diagram 7), but also the points of division around the pattern should create slightly wedge-shaped pieces (*see* diagram 3 a, b) that naturally lock together, but will release from the pattern and subsequent castings with ease. Other registration methods are 'nipples and cups' and 'pinch lines' (*see* diagram 8).

Securing Mould Pieces Together

This can be done with rubber bands, cut-up pieces of bicycle inner tube, wire or string ties, clamps or nuts and bolts. Holes for nuts and bolts should be drilled before the mould is released from the original pattern in order to maintain accurate registration when reassembling the mould. Whichever method is employed mould pieces need to be accurately and securely held together during casting to reduce seam lines.

Flanges

Flanges are probably one of the most important aspects to creating a successful mould. They are an extended width to

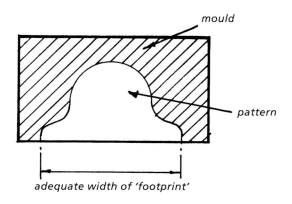

adequate width of 'footprint'

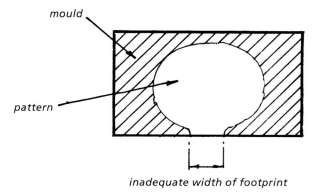

inadequate width of footprint

Diagram 5
Access into the mould

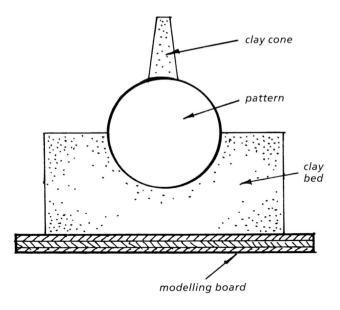

Diagram 6

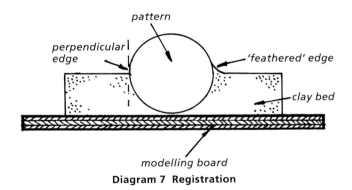

Diagram 7 Registration

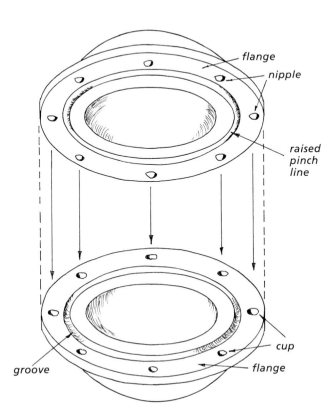

Diagram 8b Two-piece mould

Diagram 8a image labels: cup, pattern masked by clay bed, flange, cup, two-piece mould

Diagram 8a Nipple and cup registration

the edges of the mould pieces, which create a positive area of registration to locate and secure the mould pieces together (*see* diagram 8).

Walls

Patterns that are too big to lay down or cannot be moved will be a problem when building clay beds up to division lines. In these cases walls can be built onto the division lines on a pattern to mask off mould pieces as they are being made. In the same way as with a clay bed, the walls are removed and repositioned as the mould pieces are made. Walls can be made of clay, brass shim (*see* plaster waste moulds), or any other material that can be cut to shape and temporarily attached to the pattern (*see* diagram 9).

Mould Categories and Types

Moulds can be split into four principal categories, of which there can be up to four types.

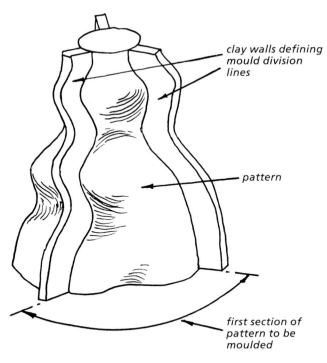

clay walls defining
mould division
lines

pattern

first section of
pattern to be
moulded

Diagram 9 Walls

Categories

- **Rigid** This type of mould is constructed from a rigid material, and will usually need to be constructed as a multi-piece mould to facilitate the removal of the rigid material around any undercutting on the pattern.
- **Flexible** These moulds can be stretched over a certain amount of undercutting on a pattern and can therefore sometimes be constructed as one-piece moulds. Patterns with deep undercutting may require a flexible mould to be made in pieces. However, due to its flexibility, the mould can be made in less pieces than a rigid mould.
- **One-off cast** These moulds will produce a one-off casting before the mould is destroyed.
- **Multiple cast** These moulds can be used to produce multiple castings until the mould becomes depleted.

Types

- **Type 1: one piece** This type of mould is constructed around a pattern in one piece (*see* diagram 10a).

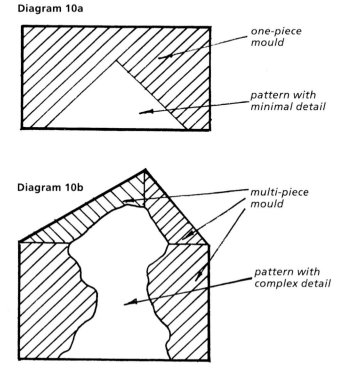

Diagram 10a

one-piece
mould

pattern with
minimal detail

Diagram 10b

multi-piece
mould

pattern with
complex detail

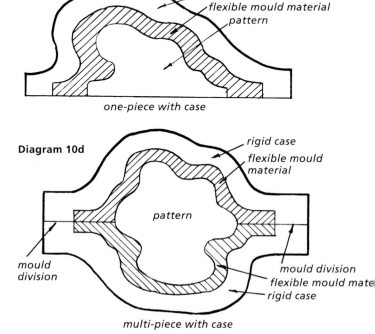

Diagram 10c

rigid case

flexible mould material

pattern

one-piece with case

Diagram 10d

rigid case

flexible mould
material

pattern

mould
division

mould division
flexible mould mate
rigid case

multi-piece with case

- ■ **Type 2: multi-piece** These moulds are constructed around a pattern in two or more pieces (*see* diagram 10b).
- ■ **Type 3: one piece with case** This is a thin, flexible mould made in one piece that is supported by a rigid case (*see* diagram 10c).
- ■ **Type 4: multi-piece with case** This is a thin, flexible mould made in two or more pieces that are supported by a rigid case (*see* diagram 10d).

Mould Descriptions and Uses

Moulds can be made of various materials, which will in turn affect the ways in which they can be used. Some moulds can be used with a variety of casting materials, whereas in other cases certain casting materials work best with certain types of mould. Sometimes only one casting material can be used in a particular mould. It will be necessary to match the considerations of the previous section with the mould description and uses described below. Following is a brief description of the various moulds, and the casting possibilities of each, that will be covered in this book.

Plaster Moulds

Casting plasters are made from fine-quality powdered gypsum, and come in different specifications for different uses. Plaster is mixed with water in proportion, then applied to a pattern. It will set hard, forming a rigid mould.

TYPES

- ■ **One-piece moulds** Suitable only for patterns with no undercutting or for flexible patterns and casts.
- ■ **Multi-piece moulds** Suitable for patterns with undercutting. However, the more complex the undercutting, the more mould pieces will be required.
- ■ **Waste moulds** Suitable for one-off cast production, as the mould is 'wasted' (destroyed) during the removal of the cast from the mould. They can be made as a one- or multi-piece mould, depending on how easily the pattern can be removed.

- ■ **Slip moulds** Used for casting clay 'slip' (liquid clay). These moulds can be made as one- or multi-piece and are the only moulds slip can be cast in.

PATTERN OPTIONS AND RELEASE AGENT REQUIREMENT
N.B. For the purposes of the lists below, whether a pattern material requires a release agent will be denoted with a Y for yes or an N for No, the specifics of which are described in Chapter 3.

- ■ clay (N)
- ■ plaster (Y)
- ■ plastics, including polyester resin and fibreglass (Y)
- ■ wood (Y)
- ■ glass and ceramics (Y)
- ■ metal (Y)
- ■ wax (N)
- ■ textiles and papers (Y)
- ■ organics (material-dependent)
- ■ rubbers (N)
- ■ cements and stone (Y)
- ■ plasticine

CAST RUN

- ■ One piece and multi-piece: ten plus, depending on casting material, application of release agents and mould care.
- ■ Waste mould: one-off.
- ■ Slip mould: fifty plus.

CASTING OPTIONS AND RELEASE AGENT REQUIREMENT

- ■ plaster (Y)
- ■ polyester resin and fibreglass (Y)
- ■ wax (Y)
- ■ clay slip (N)
- ■ pressed clay (N)
- ■ ciment fondu (N)
- ■ silicone and vinyl rubber (N)
- ■ latex (N)

Silicone Moulds

Silicone is a synthetic rubber compound. It comes as a viscous

MOULDMAKING AND CASTING PRINCIPLES

liquid that turns into a flexible rubber material at room temperature with the addition of a catalyst. Is applied to a pattern in a semi-viscous state, setting to a flexible rubber to form the mould. Please note that the silicone rubbers dealt with in this book are of the 'condensation cure' range, that are generally easier to use in terms of preparation and versatility than the 'addition cure' range.

TYPES

- **One piece** Suitable for small patterns with little/no undercutting.
- **Multi-piece** Suitable for small patterns with more undercutting.
- **One piece with a case** A thinner, more flexible mould supported by a removable rigid case. Suitable for patterns with moderate undercutting.
- **Multi-piece with a case** As above, but suitable for patterns with deep undercutting.
- **Low-melt metal alloy moulds** Can be made as any of the above types. Suitable for casting low-melt metal alloys. Requires a specific low-melt alloy silicone rubber.
- **Food-safe moulds** As above. Suitable for casting chocolate, jelly and so on for consumption. Requires a specific food-safe silicone rubber.

PATTERN OPTIONS AND RELEASE AGENT REQUIREMENT

- clay (N)
- plaster (N)
- plastic (N)
- wax (N)
- glass and ceramic (N)
- wood (Y)
- metal (N)
- paper and textiles (Y)
- organics (material-dependent)
- cements and stone (material-dependent)
- plasticine (Y)

CAST RUN

- Twenty to 100, depending on casting material and application.

CASTING OPTIONS AND RELEASE AGENT REQUIREMENT

- plaster (N)
- polyester resin and fibreglass (N)
- wax (N)
- ciment fondu (N)
- silicone rubber (Y)
- vinyl rubber (Y)
- latex (Y)
- low-melt metal alloys (N)
- food (N)

Vinyl Moulds

Vinyl plastic moulds have the properties of a flexible rubber. Vinyl plastic comes as a block of flexible rubber that is melted to a viscous liquid, which then resets to its original state, forming the mould around the pattern. One of the great advantages vinyl moulds have over the other types of mould detailed here is the fact that they can be melted down after use and reused.

TYPES

- **One piece** Suitable for small patterns with little/no undercutting.
- **Multi-piece** Suitable for small patterns with more undercutting.
- **One piece with a case** A thinner more flexible mould supported by a removable rigid case. Suitable for patterns with moderate undercutting.
- **Multi-piece with a case** As above, but suitable for patterns with deep undercutting.

PATTERN OPTIONS AND RELEASE AGENT REQUIREMENT

- clay (N)
- plaster (N)
- plastic (material-dependent)
- glass and ceramic (material-dependent)
- metal (N)
- cement and stone (N)

CAST RUN

- Ten to fifteen, depending on casting material and application.

CASTING OPTIONS AND RELEASE AGENT REQUIREMENT

- plaster (N)
- polyester resin and fibreglass (N)
- wax (N)
- ciment fondu (N)
- silicone rubber (Y)
- vinyl rubber (Y)
- latex (Y)

Life Moulds

Life moulds are made from fast-setting hypoallergenic (skin-safe) materials that are applied directly onto the skin of a life model, forming a mould of body parts. They are usually used to create a one-off plaster casting (the mould is destroyed in the process of removing the cast from the mould) that can be used to remould from if multiple castings in different materials are required. There are two moulding materials that are safe to use on the skin. One is plaster bandage, which is a fine cotton scrim (open-weave textile) that is impregnated with plaster. After dipping in water, the plaster bandage is applied to the skin and allowed to set to form a rigid mould. The other is alginate, which is a powder that, when mixed with water, is applied to the skin and allowed to set forming a flexible rubber mould. This is then backed up with a plaster bandage, which provides a rigid case to support the alginate.

TYPES

- **Multi-piece plaster bandage** A rigid mould suitable for all life-moulding purposes, producing moulds with a good degree of reproductive quality.
- **Multi-piece alginate with a plaster bandage case** A flexible mould supported by a rigid case suitable for all life-moulding purposes, producing moulds with a high degree of reproductive quality.

N.B. It is possible to combine the two types of mould to produce a combined single mould.

LIFE MODEL AND RELEASE AGENTS
It will almost certainly be necessary to use a life model when making life moulds as it is a job that requires two hands and is very difficult to perform on your own body as you would need to keep perfectly still.

The release agents required for life moulding are petroleum jelly or baby oil. These should be applied directly to the skin of the model.

PATTERN OPTIONS AND RELEASE AGENT REQUIREMENT
Although life moulding materials are designed for use on a life model, in theory they could be used on a number of different pattern materials.

Alginate
- plaster (Y)
- plastic (N)
- clay (N)
- glass and ceramic (N)
- wood (Y)
- plasticine (N)
- metal (N)
- paper and textile (Y)
- organics (material-dependent)
- wax (N)
- rubbers (N)
- cement and stone (Y)

Plaster Bandage
- plaster (Y)
- plastic (Y)
- clay (N)
- glass and ceramic (Y)
- wood (Y)
- metal (Y)
- paper and textile (Y)
- organic (material-dependent)
- wax (N)
- rubbers (N)
- cement and stone (Y)
- plasticine

CAST RUN

- Usually one-off.

CASTING OPTIONS AND RELEASE AGENT REQUIREMENT

Generally a plaster cast is taken from a life mould, which can then be remoulded to produce a more durable mould for multiple casts in different materials. However, in theory a number of different materials could be cast in a life mould, bearing in mind it is more than likely that it will only be possible to produce one cast.

Alginate
- plaster (N)
- wax (N)

Plaster Bandage
- plaster (Y)
- wax (Y)
- polyester resin and fibreglass (Y)
- silicone (Y)
- latex (Y)

Clay Press Moulds

These moulds use clay as the moulding material. Clay slabs are pressed onto the surface of the pattern; the clay is then removed, having picked up the detail from the pattern. The casting is taken directly from the clay. Alternatively, small patterns can be pressed into a block of clay; when they are removed the impression of the pattern is left in the clay block, which is then cast into directly. The natural flexibility of clay allows these moulds to cope with a certain amount of undercutting, as the clay mould can be peeled away. These moulds are generally only suitable for one-off castings, the moulds being destroyed in the process of removing the cast from the mould. They have relatively accurate detail capture, but accuracy in joining casts made in pieces can be difficult, and there can be considerable trimming and making good of seams to do.

TYPES

- **One piece (block)** A flexible mould suitable for small patterns with minimal undercutting/detail.

- **Two piece (block)** A flexible mould suitable for small patterns with more undercutting/detail.
- **One piece (slabs with plaster case)** A flexible mould supported by a rigid case suitable for large patterns with minimal undercutting/detail.
- **Multi-piece (slabs with plaster case)** A flexible mould supported by a rigid case suitable for large patterns with deep undercutting/detail.

PATTERN OPTIONS AND RELEASE AGENT REQUIREMENT

The physical techniques mean press moulding needs a pattern that will not distort or break when the clay is pressed onto it.

- plaster (Y)
- fired clay (Y)
- wood (Y)
- glass or ceramic (Y)
- metal (Y)
- cement and stone (Y)
- organics (Y)

CAST RUN

- One-off.

CASTING OPTIONS AND RELEASE AGENT REQUIREMENT

- plaster (N)
- wax (N)

Fibreglass Moulds

These moulds are formed from a combination of polyester resins and fibreglass mats applied to the surface of a pattern. Polyester resin is a viscous liquid that sets hard with the application of a catalyst; fibreglass mat is a reinforcing material to provide strength to the mould.

TYPES

- **One piece** Suitable only for patterns with no undercutting or for flexible patterns and casts.

■ **Multi-piece** Suitable for patterns with undercutting. However, the more complex the undercutting, the more mould pieces will be required.

CAST RUN

■ One hundred plus

PATTERN OPTIONS AND RELEASE AGENT REQUIREMENT

■ clay (Y)
■ plaster (Y)
■ plastics, including polyester resin and fibreglass (Y)
■ wood (Y)
■ glass and ceramics (Y)
■ metal (Y)
■ textiles and papers (Y)
■ organics (Y)
■ rubbers (N)
■ cement and stone (Y)
■ plasticine (Y)

CASTING OPTIONS AND RELEASE AGENT REQUIREMENT

■ plaster (Y)
■ polyester resin and fibreglass (Y)
■ wax (Y)
■ pressed clay (N)
■ ciment fondu (Y)
■ silicone and vinyl rubber (N)
■ latex (N)

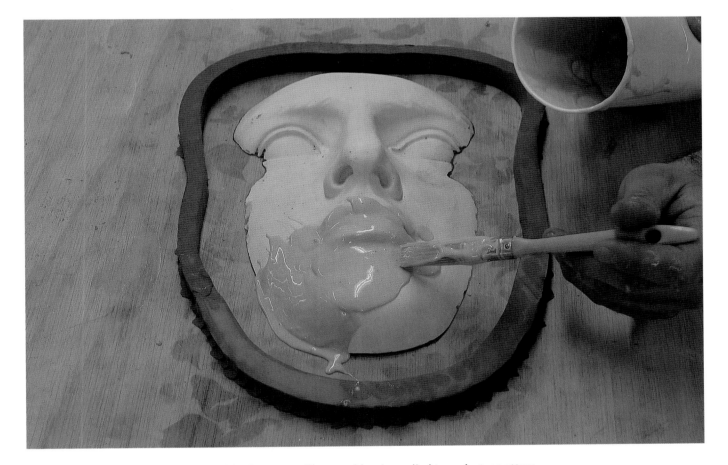

A mould is begun as silicone rubber is applied to a plaster pattern.

Applying a release agent to a mould prior to casting.

CHAPTER 3

RELEASE AGENTS

Release agents are usually essential when mouldmaking and casting, either to release moulds from original patterns or to release casts from moulds. A release agent is a substance or a combination of substances that is applied to the pattern or mould surface to act as a barrier to prevent the mould or cast sticking. Application of a release agent is also sometimes necessary while creating multi-piece moulds to prevent the pieces sticking together while they are being made. There is a wide range of release agents available for different jobs, from highly developed chemical compounds to water. Often a number of different release agents could be used for the same job; however, some may be more suitable to a specific job depending upon requirements.

Some patterns may discolour permanently with the application of certain release agents. This may be undesirable and so a range of alternatives may need to be tested on an inconspicuous part of the pattern. Similarly, some casts may be affected by release agent applications to a mould surface. The development of modern self-releasing moulding compounds has made the application of release agents obsolete in some cases, but research should always be conducted to ensure the release-agent necessity of both moulds and casts.

The question of releasing capacities of a mould may determine which moulding process is most suitable for the job required. For instance, if a large cast run is required from a mould the self-releasing properties of a rubber mould may be an advantage to avoid repeated applications of agents.

Whatever the job, a careful understanding of release agents is a requirement before embarking on any mouldmaking and casting project.

Types of Release Agents and Their Application

1. Petroleum Jelly

Petroleum jelly is a staple, all-round release agent that is efficient on all rigid moulds. It provides a greasy, waterproof barrier.

APPLICATION
It can be applied by a soft brush or by hand. It should be applied sparingly to non-porous moulds or patterns in order to avoid leaving swirls of jelly that may be picked up on the mould or cast surface. On porous surfaces it should be allowed to soak into the surface. Do not over-apply so that it clogs detail.

2. Parting Wax

Parting wax is a wax-based release agent with the consistency of a soft furniture wax. It is usually used in conjunction with shellac and PVA as a release agent for polyester resins and fibreglass, but is a useful all-rounder to release all sorts of materials.

APPLICATION
Parting wax is applied by brush or cloth. Allow it to dry for a few minutes and then buff with a dry, lint-free cloth. It should be applied two to three times for maximum releasing properties. Do not over-apply so that it clogs detail.

3. PVA Release Agent

PVA release agent is water-based, and is used in conjunction with shellac and parting wax to release polyester resins and fibreglass. PVA leaves a skin on the cast or mould surface, which is then washed away. It is either clear or has a tint of colour so that you can see where to wash it off the mould or cast.

APPLICATION

PVA is applied by brush. As it is water-based and is applied over a parting wax it is sometimes difficult to apply evenly. Keep it thoroughly mixed during application and apply more than one coat if necessary.

4. Spray Wax

This is a wax-based aerosol release agent that is used to release silicone rubbers from themselves, as with multi-piece moulds or taking silicone casts from silicone moulds. It can also be used where it is difficult to apply other release agents by hand.

APPLICATION

Spray it sparingly at a distance of 150–200mm, then reapply.

5. Shellac/Sanding Sealer

Shellac/sanding sealer is used to seal porous moulds or patterns prior to the application of a wax-based release agent. It is usually used in conjunction with parting wax and PVA to release polyester resins and fibreglass.

APPLICATION

Apply by brush sparingly and do not allow to pool into the pattern detail. Allow to dry, reapply, then allow to dry thoroughly before application of other release agents.

6. Water

Water is used to saturate plaster moulds prior to the application of wax castings. The hot wax sets instantly on contact with the saturated mould surface. It is also used to saturate porous patterns prior to vinyl mouldmaking. The water displaces and replaces the air within the pattern, essential to prevent air getting into the hot vinyl and creating pinprick holes on the surface.

APPLICATION

The mould or pattern should be fully immersed in cold water. Moulds should be soaked for one to two hours depending on their size. Patterns that are to be vinyl-moulded should be soaked for several hours.

7. Soft Soap

Soft soap is a gelatinous soap that is used to release plaster from plaster.

APPLICATION

Dissolve the soft soap with water to a form a thin but soapy liquid and apply by brush. Allow it to dry thoroughly, then apply a second application and allow to dry before use. Moulds or patterns should be wet before application. Small bubbles on the surface of patterns or moulds will disperse as the soap dries.

8. Talcum Powder

Talcum powder can be purchased in its raw, unperfumed state from mouldmaking and casting suppliers, but cosmetic talc can also be used. It is used to release clay press moulds from patterns and to release vinyl from vinyl when making piece moulds or casting vinyl in vinyl moulds.

APPLICATION

Over-apply talcum powder by brush and then brush off the excess, leaving a fine coating of powder. Moulds should be very dry and excess talc should be thoroughly brushed out of fine detail.

9. Scopas Parting Agent

This is a water-based release agent that is used to release polyester resins and fibreglass from plaster waste moulds.

APPLICATION
Apply one coat by brush onto a dry mould. Allow it to dry and use the mould within two hours of application.

10. Penetrating Spray

This is a product available at auto/DIY suppliers, and is used to assist car starting problems and to free seized nuts and bolts. However, it is a useful release agent for polyester resins and fibreglass from wet clay. Polyester resin does not set properly against a wet surface; the penetrating spray repels the moisture away from the clay surface of a pattern or clay bed, allowing the resin to set. It could also be used if taking polyester resin casts from a clay press mould.

APPLICATION
Apply several thin applications from a distance of 150–200mm. Allow to dry thoroughly between applications and before mouldmaking or casting.

11. PS8 Sealer

Porous patterns are usually soaked in water prior to casting in order to displace the air, but wood patterns would distort or crack if treated in this manner. PS8 is therefore used for displacing air within a wood pattern prior to vinyl moulding. It has the consistency and characteristics of a light cooking oil (and can be substituted with a light cooking oil); however, it is worth noting that the oily nature of the sealer may discolour certain patterns.

APPLICATION
Brush on as many coats of PS8 as necessary until the pattern is saturated.

12. Clay Wash

A clay wash is a very watery solution of any clay, that can be applied to moulds or patterns to release plaster casts or moulds. It can clog mould detail if over-applied or be inefficient if under-applied. It can discolour moulds or casts depending on the clay type.

APPLICATION
Dissolve a little clay in plenty of water and apply by soft brush. Allow to dry, then reapply. Allow to dry before moulding or casting.

13. Melamine Foil/Polythene Sheet

These materials can be used to line moulds to prevent polyester resins (and most other materials) adhering to them. They are suitable only for moulds with geometric flat planes, as they will not bend accurately into convex or concave areas. They can have the advantage of providing a very smooth mould or cast surface.

APPLICATION
Cut the sheet to fit the mould or pattern face and apply with spray adhesive. It may be necessary to rub a little parting wax along the join lines.

14. Combination (5+2+3)

This is a combination of shellac, parting wax and PVA release agent applied to a mould or pattern to release polyester resins and fibreglass. Combination (5+2+3) refers to their respective numbers in this list of release agents.

APPLICATION
Apply two coats of shellac, allowing to dry thoroughly between coats. Apply one coat of wax and allow to dry for a few minutes before buffing with a dry lint-free cloth, and repeat. Apply one coat of PVA and allow to dry thoroughly.

Tips: Do not allow the shellac to pool in the detail. Do not allow the wax to clog the detail. Keep the PVA thoroughly mixed while applying; if it will not cover fully, allow what has covered to dry thoroughly and apply a second coat. When removing casts or moulds from combination, the PVA will come away with them as a thin skin, which should then be peeled and washed away.

Application of talcum powder release agent
to a pattern prior to mouldmaking.

Pattern Release Agents (applied to pattern material to release mould)

PATTERN MATERIAL	MOULD TYPE							
	plaster	plaster waste	fibre-glass	clay press	vinyl	silicone	life (plaster bandage)	life (alginate)
clay	none	none	10	n/a	none	none	none	none
plaster	1/2/7	1/2/7	14	8	6	none	1/2/7	1
plastics	1/2	1/2	2+3	8	none*	none	1/2	none
wood	1/2	1/2	14	8	n/a (11)*****	2	1/2	1/2
glass/ceramic	1/2	n/a	2+3	8/1/2	none*	none	1/2	none
metal	1/2	1/2	2+3	8/1/2	none	none	1/2	none
wax	none	none	n/a	n/a	n/a	none	none	none
textile/paper	1/2/4/7	1/2/4	14	8	n/a	1/2/4/7	1/4/4/7	1/2/4/7
organic (waxy/oily)	none	none	2+3	none**	n/a***	none	none	none
organic (dry/hairy)	1/2	1/2	14	1/2*	n/a***	1/2	1/2	1/2
rubber	none	none****	none	none**	none	4	none	none
cement/stone	1/2/7	n/a	14	8	6	1/2	1/2	1/2
plasticine	5+2/1	5+2/1	14	n/a	n/a	5+2/1	5+2/1	4
life model	n/a	n/a	n/a	n/a	n/a	n/a	1	1

*   test plastics will tolerate vinyl temperature

**   dependent on whether material will distort or damage while being 'pressed'

***   unless organics are robust they may split due to heat

****   as long as rubber pattern is strong enough to release from mould without damaging it

*****   it is sometimes possible to soak wood patterns in PS8/oil to displace air. However, this may not always be desirable if it permanently discolours original pattern

Mould Release Agents (applied to mould to release cast)

CAST MATERIAL	MOULD TYPE							
	plaster	plaster waste	fibre-glass	clay press	vinyl	silicone	life (plaster bandage)	life (alginate)
clay (pressed)	none	n/a	8	n/a	none	none	n/a	n/a
plaster	1/2/7	7/12	1/2	none	none	none	1/2	none
polyester resins	14/9	14/9	2+3	10	none	none	14	10
ciment fondu	1/2/7	1/2/7	1/2	n/a	none	none	n/a	n/a
silicone rubber	none	none	none	none	4	4	none	none
vinyl rubber	6	6	n/a	n/a	8	4	n/a	n/a
wax	6	n/a	n/a	none	none	none	n/a	n/a

Numbers correspond to the numbered release agents described in the preceding pages.

Four-piece plaster mould.

CHAPTER 4

PLASTER MOULDS

Plaster is a versatile and accurate mouldmaking medium. It is also relatively cheap and fast to use compared with most other moulding processes. Casting plaster is a gypsum-based material that reconstitutes to a hard, durable medium when water is added in the correct quantities. Because of the firm (non-flexible) nature of set casting plaster, it is usually necessary to create plaster moulds in several parts in order to get around any undercutting problems on the pattern and to allow subsequent castings to be removed from the mould. This creates a challenge for the mouldmaker that the more modern flexible casting materials have been designed to overcome. The problems of creating multi-piece non-flexible moulds should not be underestimated – such moulds need to be planned and created with accuracy and care in order to achieve moulds that work efficiently during multi-cast production. It may therefore be unacceptable to create multi-piece moulds from very complicated originals in terms of time and labour. Waste moulding, in which the mould is chipped away from the pattern and not reused, may be a solution to the labour involved in multi-piece moulding.

Another consideration should be the type of casting plaster used. This will largely depend on the size and type of mould to be made and what casts are to be made from the mould. Economy of material is another consideration.

Whether creating waste, one-piece or multi-piece moulds, plaster is a traditional, accurate and challenging material and will certainly educate the mouldmaker in all the fundamental principles of mouldmaking. Specifications regarding the material, tools and principal techniques of using plaster should be read before embarking on mouldmaking (*see* Chapter 2).

Plaster moulds can be taken from a number of different pattern materials, the type of which will determine whether or not a release agent will be required. If making multi-piece moulds, pattern undercutting, mould division lines and the method of masking the divisions (clay bed or walls) will need to be determined. These considerations are also common to other rigid moulding processes (*see* Chapters 2 and 10 for plaster-working principles).

Waste Moulds

Waste moulding is one of the most cost-effective and quickest methods of mouldmaking. The basic principles are to create a plaster mould around a soft pattern material, remove the soft pattern, fill the mould with a durable casting material and chip away the mould to reveal the cast.

As the name suggests, waste moulds can only be used once, the mould being destroyed in the process of removing the cast from the mould; consequently, these moulds are not suitable for multi-cast production. They are generally used to replicate quickly a pattern created in a soft material into a more durable material, usually plaster. If multiple casts are later required, the mouldmaker can then remould from this cast in a more suitable moulding material.

The mould is created by building up layers of plaster directly onto the surface of the original. The first layer of plaster applied to the surface is coloured in order to give a warning that the mouldmaker is getting close to the surface of the cast material when chipping away the mould from the cast.

Obviously, because of the single-use nature of these moulds they will not be suitable for all mouldmaking jobs, and this will need to be considered. The other consideration is that the original material should be soft (for example clay, plasticine, Chavant clay), in order for it to be removed easily without incurring damage. When materials other than clay are used consideration should be given to whether or not they require a release agent (*see* Chapter 3).

Waste moulds can be created around hard pattern materials, but would need to be multi-piece in order to be removable from the pattern, thereby compromising the main point of waste moulding, which is to create a cheap, quick and easy mould without the need for too much consideration of the problem of undercutting.

The only time when consideration needs to be given to undercutting while making waste moulds is when it would be difficult to remove the pattern material. An example of this would be when making a waste mould of a portrait head, where the size of the neck aperture, which will be the size of the mould mouth, is too small for the removal of the original material and filling with the casting material. In this case, the mould has to be made in more than one piece to allow for access. Similarly, if a pattern has a footprint (base) too small to create a large enough mould mouth, the mould will need to be made in more than one piece (*see* Chapter 2 for moulding principles).

One-Piece Waste Moulds

METHOD

1. Create a wall of clay around the foot of the pattern to retain the plaster as it is applied. This should be 30mm in height and at a distance of 30mm from the foot of the original.
2. Treat the area of modelling board between the inside of the wall and the original with a release agent so that when the plaster is applied it will not stick.
3. Apply the first plaster layer of the mould. The principal technique for plaster application is the same as for making a hollow cast from a mould (*see* 'Plaster Casting'). The first plaster mix should be to the consistency of double cream, with enough colour added to tint it slightly. Any water-based paint or pigment can be used, for example children's pre-mixed poster paint. This will act as the warning coat when chipping off the mould from the cast. This initial layer should be applied methodically and carefully as it is the layer that picks up the detail of the pattern and will become the casting surface of the mould.

   There are a number of different methods for application of plaster in layers, but brush application is the most efficient and easy to control. The first layer should be applied evenly to a depth of approximately 2–5mm depending on any undercutting on the pattern. In areas of deep undercutting the plaster layer should be blown directly after application,

as this will force the plaster into the undercut and ensure that the plaster thoroughly covers the surface of the pattern. Allow the first coat to dry until firm to the touch. It is important not to let the first layer dry completely, in order to get good lamination between layers. It should be left to dry just long enough so that subsequent layers will not break through when applied.

4. Apply subsequent layers batch by batch until a total mould thickness of approximately 18mm is achieved. The total thickness may vary depending on the size of the mould, and larger moulds may need to be thicker and be reinforced with scrim or fibreglass, but bear in mind that a waste mould will need to be chipped away from the casting and any reinforcement will make this considerably more difficult.
5. Once total thickness has been achieved, leave the mould to set thoroughly. This will take between two to three hours, after which the pattern can be removed.

OPENING AND CLEANING

1. Remove the retaining clay wall and release the whole mould, containing the pattern, from the modelling board. Invert to reveal the opening to the mould.
2. Carefully remove the pattern from the mould so as not to damage the mould surface. The pattern can initially be removed in large pieces by hand, and then carefully picked out with clay tools where undercutting occurs. Use wood or plastic clay tools so as not damage the mould surface.
3. Wash the mould thoroughly to remove any last bits of the pattern material. Once emptied and clean, the mould is ready for casting and release agents can be considered.

RELEASE AGENTS AND CASTING

When casting from a waste mould, as with any mould, consideration should be given to release agents that will need to be applied to stop the cast from sticking to the mould. Theoretically, any casting material could be used a waste mould, as long as the correct release agent is applied and the chosen material can withstand the rigours of the chipping-out process (*see* below).

If casting in plaster, the mould can be used as soon as it is set, even if it is still wet (in fact, it is preferable to have the mould damp); apply soft soap liberally in two or three coats and allow to dry. When casting in other materials, for example polyester resins, the mould will need to be allowed to dry thoroughly and the correct sequence of agents applied (*see* Chapter 3). Once the release agent has been applied, the mould is ready to cast (*see* Chapter 2 and relevant 'Casting' chapter).

CHIPPING OUT THE CAST

Once the cast, of whichever material, has been allowed to set thoroughly the chipping-out procedure can commence. This requires the chipping away of the mould bit by bit to reveal the cast. The best tools for this job are old wood chisels, stone carving chisels or old screwdrivers and a wooden mallet of a size that feels comfortable to the mouldmaker and the size of the chisels. The size of the tools will depend on the size of the mould, but a variety of sizes is useful for different stages of the operation.

1. Place the mould on a piece of damp, folded cloth to stop slippage and to absorb some of the impact from the mallet blows. Initially, use the larger chisels, starting from an area at the top of the mould. Where possible, hold the chisel at a slight angle to lessen the impact of the blows to the cast. Working over the whole surface of the mould, remove the plaster down to the coloured warning layer.
2. Once flashes of the warning colour are observed, proceed with the smaller chisels and a little more caution, as the surface of the cast will be near. Remove a small area, 25–50mm, of the coloured layer first to reveal the surface of the cast, expanding this slowly to reveal more and more of the cast. It may be necessary at this stage to use very small tools (dental tools are good if you can get them), gently levering pieces of the coloured layer away. If necessary, apply hot, soapy water to aid removal of the mould.

   Particular care should be taken at points where the cast is protruding and where there may be a lot of surface detail, but however careful or experienced the mouldmaker, there will inevitably be the odd slip during the chipping-out process.
3. Once chipping out has been completed, wash the cast thoroughly and make any necessary repairs (*see* the next section for the photo sequence of this procedure).

Multi-Piece Waste Moulds

Any pattern without a wide enough footprint or base to allow removal of the pattern material through the mould mouth will require a mould made in more than one piece. Multi-piece moulds can be made around patterns that are completely in the round, that is, they have no footprint. However, consideration will need to be given to allow an opening into the mould through which the casting material can be introduced. A pattern with a footprint will have the mould constructed around it, leaving an opening into the mould; patterns in the round will need an opening created in one of the pieces to allow access. This is done by attaching a temporary stalk to the body of the pattern, around which the mould is made. When the mould is complete and the stalk removed, an opening into the mould is left.

When making any multi-piece mould, the division of the mould pieces should occur along the high points of any undercut areas, so that they can be released from the pattern. The division lines should be determined and either marked on the pattern surface or committed to memory.

The next consideration will be the method of masking each area of the pattern as it is made. Each piece of the mould will need to be made separately and the other areas of the pattern that the mould pieces are to be made from will need to be masked while each piece is made. This can either be done with a temporary clay bed or walls that are built up to the division lines. The mould piece is then made, the bed or wall removed and reconstructed along the next set of division lines defining the next piece to be made.

There are generally two methods employed for creating the division walls for a multi-piece waste mould, brass shim and clay.

BRASS SHIM DIVISION WALLS

Brass shim is very thin brass sheeting (0.004in) that usually comes in a roll and can be cut with scissors and pushed into the soft pattern material creating a wall.

1. Once the division lines on the pattern have been defined it is a good idea to score lightly the surface along these lines as a guide for the shim wall to be followed. It is preferable to 'wave' the division lines in order to provide registration between mould pieces.
2. Cut enough brass shim into pieces to create the walls that will define the first section of the mould. The walls should be slightly angled to varying degrees, and of approximately 30mm in width (*see* diagram 1). The lengths of the pieces should be varied to accommodate undulating profiles along the division lines and the size of the pattern.

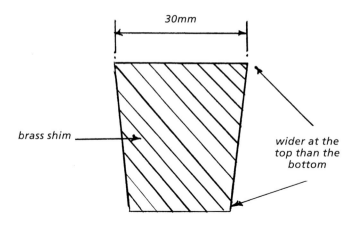

30mm

brass shim

wider at the
top than the
bottom

Diagram 1. Brass shim piece for creating division walls

3. Push the shim approximately 12mm into the surface of the pattern along the division lines, leaving 30mm exposed to create the wall. Each piece of shim should overlap the next by approximately 1–2mm, and should be set perpendicular to the surface of the pattern.
4. As the shim is so thin, it can be curved while pushing in to follow the direction of the line. At intervals along the wall it is a good idea to bend a piece of shim into a point, so as to create registration between mould pieces.

   If moulding a standing pattern that has a base, the shim walls will need to be taken down to the modelling board for each section. The pieces of the mould will be built against the modelling board at these points, creating the mouth of the mould through which it will be filled. If moulding a pattern completely in the round, a stalk of clay should be attached to the pattern. When the mould is complete and the stalk removed it will leave a hole in this piece of the mould, through which the casting material can be poured. Stalks should be positioned on a section of the pattern where the subsequent casting can be easily reworked, as there will be a loss of detail at this point. Stalk diameter will depend on the size of the pattern and what casting material will be poured into the mould.
5. Once the shim wall has been created, the first piece of the mould can be made, following the method for one-piece moulds above, steps 3–6. The mould should be built up the walls at least 50mm to provide a flange around the mould piece. The first piece of the mould should be allowed to set until firm enough not to break up when removing the shim.

6. Remove the shim carefully all the way around the first piece, exposing the plaster.
7. Now the next set of shim walls can be created to define the next adjoining piece. One section of the wall defining the second piece will be a section of the first plaster piece, creating the connection between the mould pieces. A release agent will need to be used along this section so that the pieces can be separated once the mould is complete.
8. The second piece can now be built up in the same way as the first, and the whole procedure repeated for as many pieces as the mould requires, each piece connecting to the previously made piece until the mould is complete.
9. Once all the mould pieces have been made the mould can be released from the original.

OPENING, TRIMMING AND CLEANING

Before opening the mould it is important to consider how the pieces of the mould will be held together while casting. With the smaller moulds it may be adequate to use large rubber bands or cut-up lengths of bicycle inner tube, tied and made into bands of appropriate lengths. For the larger moulds the mould pieces can be bolted together – this is where a wide flange is vital. A wide flange will provide a large surface area of registration between mould pieces, but also plenty of space to drill holes to take a nut and bolt. If using nuts and bolts to secure mould pieces, all the holes to take them should be drilled before opening the mould to ensure accurate reassembly. Any hand or electric drill with a standard HSS (high-speed steel) twist drill bit can be used to do this. Bolts with wing nuts should be used for ease of reassembly, and a large washer on each side of the bolt (all should be galvanized so as not to rust).

The mould can now be opened, trimmed and cleaned.

1. Cut back all flanges by a few millimetres to expose the seam between them and provide a strong, tidy edge. This can be done with a surform, a rasp-like tool that is very effective for removing large quantities of material. Take the excess plaster back until a line of the first coloured layer can be seen between the mould pieces. With all of the seams exposed, the mould pieces can be removed.
2. Start by inserting a thin, strong-blade knife in-between all the mould pieces. Initially, this is just to part the adjoining flange seams of each piece, not to remove the pieces from the pattern.

MULTI-PIECE WASTE MOULD

The original clay pattern.

The shim wall curves around the ear creating a 'wave'.
A clay retaining collar curves around the neck.

Inserting the brass shim.

Apply the first colour layer by brush and blow into the fine
pattern detail.

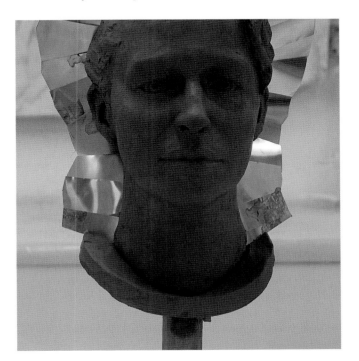

MULTI-PIECE WASTE MOULD *continued*

Application of the secondary back laminate layer.

With the second piece complete, the two halves can be separated and the clay removed ready for casting.

With the first piece complete, remove the shim ready to begin the second piece.

Chipping away the mould after plaster casting.

The cast emerges from the waste of the mould.

3. Once the flanges have been parted the seams can be opened up more by methodically working around the seams. Open the seams equally and gradually all the way along, rather than in one place.
4. As the seams start to open it may be necessary to use a chisel, particularly on larger moulds. Larger moulds may benefit from a series of wooden or plastic wedges being gradually driven into the seam in sequence until the mould pieces separate. Once one mould piece has been removed the others will generally come away easily.
5. With all the mould pieces removed from the pattern, sand the flange edges and the outside of the mould with abrasive paper for ease of mould handling. Do not work the inside of the mould as it will compromise the moulded surface.
6. Wash the mould in warm water and detergent, and dry it by hand.
7. Reassemble the mould securely to prevent it warping before it is used.

N.B. Before chipping out a multi-piece waste mould remove all nut and bolts or ties securing the pieces.

CLAY DIVISION WALLS

Another method of creating mould piece divisions while building a multi-piece mould is to create a clay wall.

1. Strips of clay approximately 6mm thick, 18mm wide and 150mm long should be prepared using a clay harp or rolling pin. It is a good idea to use clay of a different colour to the pattern material to provide some definition, and also of a slightly harder consistency so that it does not merge with the pattern material.
2. Create division lines between mould pieces as outlined in the brass shim wall method above, but using clay. The clay wall should be applied firmly at right angles to the surface of the pattern. It should not be smoothed into the surface so as to avoid feathering (weak inner edges between mould pieces), but should be applied firmly enough to stay. Quite long sections of the wall can be built on at a time to avoid too much joining. Where sections do join, the clay should be smoothed to create a smooth, continuous wall.
3. If necessary, build buttresses from behind to support the walls. At any point of wall contact, the pattern may be marked or damaged. This can be cleaned up after removal of the walls between making the mould pieces.
4. As with a brass shim wall, it is a good idea when deciding on division lines between parts to undulate the line to create good registration. However, it may be necessary to have some areas of more positive registration. These can be created using any small tool that has a handle with a rounded end that can be pressed into the wall at intervals, making a shallow depression that will create a more positive 'nipple and cup' registration between mould pieces.
5. Follow the one-piece mould method steps 3–6. The mould should be built up the walls at least 50mm to provide a flange around the mould piece.
6. Repair any damage to the pattern as you go.
7. It is sometimes a good idea to build in clay wedges onto the plaster wall between pieces to allow some point of purchase and water access between mould pieces when demoulding.
8. Once the mould is complete, opening, trimming and cleaning procedures can be carried out as for the brass shim wall method above.

N.B. One other method of creating the divisions while waste mouldmaking is the clay bed procedure, which will be outlined below under 'Reusable Plaster Moulds'.

Reusable Plaster Moulds

The previous section dealt with one-use moulds that are destroyed during removal from the cast and cannot be used again. In this section, we shall be dealing with plaster moulds that can be reused. Casting plaster, or gypsum, will pick up very fine detail; it is also relatively durable, and therefore as long as it is handled carefully it will survive many castings.

For most casting materials, plaster moulds of whatever description will need to be thoroughly primed with the appropriate release agents each time the mould is used. This done, many casts can be produced from the same mould before any loss of detail is distinguishable.

Due to the rigid nature of set plaster, however, the utmost care must be taken to avoid undercutting within the mould to ensure that castings can be removed without damage. When making plaster waste moulds as described previously a certain amount of undercutting can be ignored as long as the original material can be removed, because the mould is eventually chipped away and wasted. The high points and points on the original pattern where there is undercutting must be very carefully planned before embarking on reusable plaster moulds, thus enabling the mould to be removed without damaging the castings. Reusable plaster moulds will usually need to be made in several pieces in order to overcome these problems, although it is sometimes possible to make a reusable plaster mould in one piece (*see* Chapter 2 for mouldmaking principles).

One-Piece Reusable Plaster Moulds

One-piece plaster moulds are only possible when the pattern is of a shape with no undercutting, so that the cast material can be removed without damage to it or the mould. This means that the footprint (base) of the pattern by definition will need to be wider than the top (for example, conical or pyramidal). The footprint of the pattern in its negative moulded form will become the mouth of the mould, where it is filled with the casting material.

The casting material should also be considered, as certain materials will come out of a plaster mould easier than others. For instance, if the mould is being used to cast clay or wax, both of these materials shrink slightly, making it easier for the cast to drop out of the mould. There is very little shrinkage in a plaster cast, however, potentially making it very difficult to remove, even if the pattern has no undercutting.

Remember that if the pattern is of a soft material it may be possible to remove it from the mould, but if subsequent casts are of a more durable material they may be impossible to remove if any undercutting is present.

These problems considered, one-piece plaster moulds can be made in two ways, block or laminated. Block moulds are made by pouring blocks of plaster and should therefore only be considered for relatively small patterns, due to the volume and subsequent weight of plaster used to make them. Laminate moulds can be made much lighter, as they are made by building up layers of plaster and laminate, and should be used for larger patterns.

BLOCK MOULDS

1. Secure the pattern firmly to a modelling board and fill any gaps between the pattern's base and the board with clay to prevent plaster from seeping underneath.
2. Build a clay wall 10mm thick and 18–25mm higher than the highest point of the pattern to retain the plaster. The wall can either be built square around the pattern, or can follow the outline of the pattern. Build the wall 18–25mm from the pattern to economize on plaster. If necessary, add buttresses to the outside of the walls for extra strength. Other materials can be used to create walls (wood, card, perspex) as long as they will contain the liquid plaster while it sets. Hard materials used to create walls will need to be joined securely and sealed to the modelling board with clay to avoid leaks.
3. Consideration should be given to whether the wall and pattern material will require a release agent (*see* Chapter 3). The area of modelling board between the pattern and the inside of the wall will need a release agent.
4. Once the walls are in place and secure the mould is ready to be poured. Mix enough plaster to cover the pattern in one operation, to avoid trapping air bubbles. A standard plaster mix to the double-cream consistency should be made, avoiding as far as possible trapping air while mixing (*see* Chapter 10).
5. Pour the mix within the walls of the mould from one point in a steady continuous flow, ideally not directly onto the pattern. The plaster should be allowed to rise steadily from one point within the walls, allowing the mix to flow into all the pattern detail and push out any trapped air as it rises. The mix should be poured to the top of the mould walls, bringing it to 18–25mm over the highest point of the pattern.
6. Pouring plaster, particularly in more than one mix, will inevitably trap a little air, which should be removed as much

as possible. This can be done either by agitating or tapping the mould repeatedly, or by introducing the fingertips approximately 10mm into the surface of the plaster and moving them up and down rapidly until air bubbles can be seen appearing on the surface.

7. Leave the mould to set fully before opening.

LAMINATE MOULDS

1. Patterns should be prepared and secured in exactly the same way as for block moulds, the only difference being that the walls surrounding the original will only need to be approximately 18mm in height.

2. There are several different methods for the application of plaster in layers, but brush application is the most efficient and easy to control. The first layer should be applied evenly to a depth of approximately 2–3mm. The first coat should be allowed to dry until firm to the touch, but it is important not to let it dry completely, in order to achieve good lamination between layers. It should be left to dry just long enough so that subsequent layers will not break through when applied.

3. Apply subsequent layers batch by batch until a total mould thickness of approximately 8–16mm is achieved. The total thickness at this stage will depend on the size of the mould, and the type of plaster being used.

4. After the first few layers, the mould needs to be laminated. Hessian scrim or fibreglass mat can be used to add the necessary integral strength. The traditional scrim material is an open-weave hessian that is available in rolls approximately 50mm to 1m wide. Enough strips of scrim should be cut and prepared to cover completely the layers of plaster. Run the scrim through a bowl of plaster, making sure it is saturated, then apply it to the cast surface. Pat each piece down firmly to remove air bubbles and ensure a good lamination. If using fibreglass, use 1½oz chopped strand and split to produce a more open weave.

5. Continue laying pieces of laminate, overlapping each by a few mm until the whole surface is covered. One layer of laminate material should be enough, but on particularly large moulds a second may be necessary.

6. Once the laminate layer has been laid, apply a last layer of plaster all over, to sandwich the laminate. This can be done with a slightly thicker mix of plaster, and applied by brush or hand to make up the final thickness of plaster.

7. The mould is now complete and should be allowed to set fully (*see* photo sequences for one- and multi-piece plaster moulds and Chapter 10 for additional information).

N.B. Although a lengthy procedure, with this method a thin mould can be produced, which may be preferable in terms of economy of materials and weight of mould.

ONE-PIECE REUSABLE MOULD

The original pattern made of MDF.

A release agent is applied to the pattern and modelling board.

ONE-PIECE REUSABLE MOULD *continued*

A clay retaining wall is built and the first layers of plaster are applied.

A layer of Hessian scrim is applied.

A final layer of plaster is applied.

The finished mould is released from the pattern.

The mould is trimmed and then cleaned.

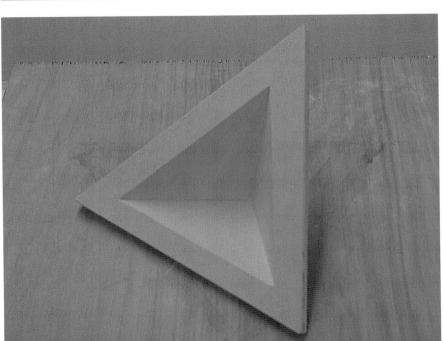

The finished mould.

OPENING, TRIMMING AND CLEANING

1. Remove the retaining walls.
2. Insert a thin, strong-bladed knife between the mould and modelling board and methodically release all the way round.
3. Carefully lever the mould off the modelling board. Patterns of a soft material will come out of the mould relatively easily; patterns of a more durable material will be more difficult, due to the suction created in the mould. A combination of tapping the mould and levering the pattern out should release it from the mould. If the pattern will not release, soaking in hot soapy water or under a running warm tap should help. Bear in mind that the pattern may be damaged during these procedures depending on what it is made from.
4. Once the pattern has been released, trim the mould. With a surform take away any sharp edges on the outside of the mould that may make it difficult to handle.
5. Wash the mould thoroughly with warm water and detergent. It is now ready for casting.

Multi-Piece Reusable Plaster Moulds

Taking plaster moulds from patterns that are 'in the round' (that is, they have no footprint or base), or with detail that is undercut will require a mould that is in several pieces in order to be able to remove the mould from the pattern and to remove subsequent casts from the mould. Plaster moulds that are in several pieces can be reused for multiple castings as long as the pieces of the mould are carefully created around any undercutting and the mould is treated with the appropriate release agents when casting. On smaller patterns the plaster pieces of the mould can be poured in blocks; on larger patterns the pieces of the mould will need to be laminated.

Clay beds or walls will need to be created along the lines of high points of undercutting on the pattern, which will define the mould pieces. The definition of these lines is the crucial point in any moulding process when the pattern is undercut, and the mouldmaker should spend some time determining where these lines should be made and mark them on the pattern if possible. Remember it may be possible to remove a pattern made in a soft material from the finished mould, but impossible to remove subsequent castings from a mould that is undercut.

Once the location of the mould pieces has been defined and marked on the pattern, the pieces of the mould need to be masked off from each other as the mould is created. This can be achieved either by defining the lines with a clay bed or with walls.

CLAY BED DIVISION LINES

1. Decide which piece of the mould is to be created first and secure the pattern temporarily to a modelling board or workbench with this section facing up as horizontally as possible (it may be easier to lay the pattern down horizontally to do this).
2. Secure with clay, bricks or pieces of wood, depending on the size of the original, wedged around the pattern to keep it in place.
3. Build up small pieces of clay to approximately 20mm below the lines defining the first piece of the mould. The bed should extend 150mm around the original. It is important to give yourself plenty of width when creating a clay bed in order to make the flanges for laminated moulds that will secure the mould pieces together. Initially, the bed can be created fairly roughly, gradually building up to the division lines with more care.
4. When approaching the division lines the bed should be smoothed off, creating a smooth surface. Care should be taken to avoid curving the clay up at the final point of contact between the clay bed and the pattern, as this will create feathering (weak inner edges between mould pieces). A clay tool should be used at this stage to create the bed perpendicular to the pattern. The bed may undulate following the lines of division. It is advisable to create the sections of the mould so that the pieces are slightly wedge-shaped to ease removal when casting.
5. Once the clay bed is firmly in place, exposing just the first piece of the mould to be created and masking off the rest of the pattern, release agents can be applied and the first piece can be made.

N.B. If the pattern is small the mould can be made up in blocks, but larger moulds will need to be laminated.

BLOCK MOULDS

To create block moulds a clay wall has to be put in place to contain the plaster as it is poured over the pattern.

1. Cut a slab of clay approximately 18mm thick and to a width 20–25mm higher than the exposed section of the pattern. The wall should be built onto the clay bed at a distance of approximately 30mm out from the division line on the exposed section of the pattern, and pinched around the outside edge onto the clay bed to prevent any leaks. Clay buttresses may also need to be built along the outside of the wall to reinforce it.

2. Make shallow depressions at intervals around the pattern into the surface of the clay bed with a round-ended tool handle to provide registration between the mould pieces. Be aware when making registration points between pieces that if the pieces cannot be taken apart vertically from one another they will not come apart (see diagrams 8a and b Chapter 2).

3. Plaster can now be poured within the clay walls. Pour a standard plaster mix steadily within the walls, in a corner rather than directly onto the pattern. Let the plaster rise steadily, pushing out any air bubbles as it rises to the top of the walls. Once poured, introduce fingertips into the surface of the plaster and move them vigorously up and down while keeping them under the surface; this will bring up any air trapped in the plaster as it was poured.

4. Leave the plaster to set fully before moving onto the next section of the mould.

5. Being careful not to remove the first piece of the mould created, remove the pattern from the clay bed and place it in a position that exposes the next section to be moulded.

6. The next clay bed can be put in place following the next section of division lines, and a clay wall built up as before. Part of the wall defining this next section will incorporate a section of the first mould piece created. The mould pieces connecting directly at this point will require a release agent to stop them sticking together (see photo sequence for 'multi-piece reusable plaster mould', pp. 46–9, for illustration in principle).

7. Pour plaster as before, creating the next piece of the mould.

8. By methodically moving from one section of the pattern to the next, creating each piece of the mould connecting to the last, all the pieces will eventually connect, completing the mould.

9. Remember when creating moulds on patterns that are completely in the round an opening into the mould will need to be made to allow access for the casting material. This should be done by placing a truncated cone (wider at the base than the top) of clay directly onto the surface of the pattern, which when removed from the completed mould will create a pouring hole into the mould at this point. It is worth considering where to place the pouring hole, as it will inevitably make a blemish on the surface of castings at this point, so put it in an area where little detail will be lost. The size of the pouring hole into a mould should be minimal and will depend to certain degree on the casting material being used in the mould, although the more viscous materials may require a larger pouring hole. If the pattern has a base this will become the mouth of the mould. If the mould is being made on a pattern with a base while it is standing the mould pieces will be made down to the modelling board or bench, leaving the underside of the base unmoulded, which will subsequently become the mould mouth. If the pattern has been laid down horizontally while moulding, a board should be placed up against the base as a surface to build the mould up against. The board will need to be repositioned as the pattern is repositioned while different parts of the mould are created (see diagram 3, Chapter 7 for principle).

LAMINATING MOULDS

Making a block mould from large patterns would require large volumes of plaster to be poured, which would be problematic to contain while setting. Moulds created from large patterns therefore require the plaster to be laminated in layers for economy of material, and subsequent ease of handling of the mould when complete.

1. Set up a clay bed as for block moulds, but the walls built onto the clay bed should be between 10–18mm high, and at a distance of 60–80mm from the pattern.

2. Lay up plaster gradually in layers to an even thickness over the pattern and clay bed. The plaster that is built up on the clay bed will become a flange surrounding that piece of the mould, and will connect to the next piece of the mould to be made. Follow the laminating method outlined in 'One-Piece Reusable Plaster Moulds', steps 2–6.

3. Each piece of the mould is made methodically around the pattern as with the block mould, each piece connecting to the previous piece until the mould is complete.

4. A pouring point into the mould will need to be considered. Follow step 9 in 'Block Moulds' above.

5. Once the mould is completed it should be left to set fully before attempting to remove it from the pattern (this will depend on which type of casting plaster has been used).

MULTI-PIECE REUSABLE PLASTER MOULD

The original pattern.

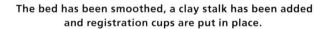

Putting the clay bed in place.

The bed has been smoothed, a clay stalk has been added
and registration cups are put in place.

The clay retaining wall is put in place.

The first layer of plaster is applied.

Hessian scrim is laminated for reinforcement.

Once the first piece has set the mould is inverted.

The clay bed is removed.

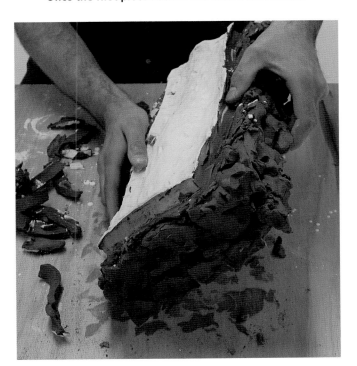

MULTI-PIECE REUSABLE PLASTER MOULD *continued*

The next clay bed is put in place.

The release agent is applied to the exposed area of the first mould piece and the next piece is begun.

The mould is set up ready for the next section to be made.

A detail showing the clay bed between two already completed mould pieces.

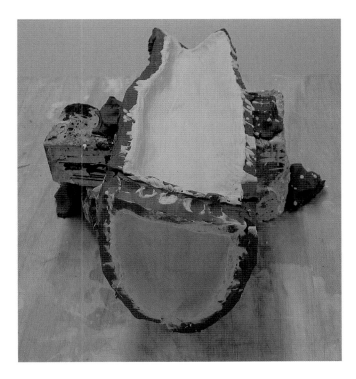

The end mould piece covering the bottle base.

Drill holes to take nuts and bolts to secure the mould pieces before releasing the mould from the pattern.

The mould pieces released from the pattern.

The mould pieces secured with nuts and bolts.

OPENING, TRIMMING AND CLEANING

Before opening the mould it is important to consider how the pieces of the mould will be held together while casting. With the smaller moulds it may be adequate to use large rubber bands or cut-up lengths of bicycle inner tube, tied and made into bands of appropriate lengths. For the larger moulds the mould pieces can be bolted together – this is where a wide flange is vital. A wide flange will provide a large surface area of registration between mould pieces, but also plenty of space to drill holes to take a nut and bolt. If using nuts and bolts to secure mould pieces, all the holes to take them should be drilled before opening the mould to ensure accurate reassembly. Any hand or electric drill with a standard HSS twist drill bit can be used to do this. Bolts with wing nuts should be used for ease of reassembly, and a large washer on each side of the bolt (all should be galvanized so as not to rust).

The mould can now be opened, trimmed and cleaned.

1. Cut back all flanges by a few millimetres to expose the seam between them and provide a strong, tidy edge. Using a sur-form, take the excess plaster back until a distinct line can be seen between the mould pieces. Once all the seams are exposed, the mould pieces can be removed.

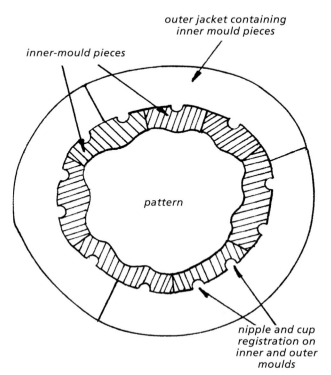

outer jacket containing inner mould pieces

inner-mould pieces

pattern

nipple and cup registration on inner and outer moulds

Diagram 2. Mould with additional jacket

2. Start by inserting a thin, strong-bladed knife in-between all the mould pieces. Initially, this is just to part the adjoining flange seams of each piece, not to remove the pieces from the pattern.

3. Once the flanges have been parted the seams can be opened up more by methodically working along them.

4. As the seams start to open it may be necessary to use a chisel, particularly on larger moulds. Larger moulds may also benefit from a series of wooden or plastic wedges being gradually driven into the seam in sequence until the mould pieces separate. Once one mould piece has been removed the others will generally come away easily.

5. With all the mould pieces removed from the pattern, sand the flange edges and the outside of the mould with abrasive paper for ease of mould handling. Do not work the inside of the mould as it will compromise the moulded surface.

6. Wash the mould in warm water and detergent, and dry it by hand.

7. The mould should now be reassembled securely to prevent it warping until it is used.

MOULDS WITH AN ADDITIONAL JACKET

Making plaster moulds from patterns that are highly detailed with a large amount of undercutting will require moulds to be made in many pieces in order to be able to remove them around the undercutting. If the mould is to be made in many pieces it may be advantageous to make a second jacket that contains the smaller pieces of the mould within (*see* Chapter 2 for mouldmaking principles).

1. As many small pieces as required should be made as described previously (if laminating the plaster it is not necessary to create flanges). The pieces should completely cover the pattern to approximately the same outside surface level, which should be relatively smooth. Ideally, the outside shape of the mould at this point should be made up to a shape that will require the containing jacket to be made in the minimum amount of pieces.

2. Depending on the shape of this inner mould it may be necessary to create 'nipple and cup' registration keys into each mould piece that will locate into the outer jacket. These can be created using a coin to scribe a cup into the outside surface of the mould. When the jacket is made the nipple will be created by the plaster flowing into the cup.

3. The seams between all the pieces should then be fully exposed using a surform or rasp and any surface coarseness

removed over the entire surface. Thoroughly treat the entire surface with a release agent.

4. Determine the division lines of high points on the mould to decide where the jacket will be divided. Either a clay bed or wall can now be built upon this line, and the first half of the containing jacket poured or laminated. Allow to set fully.

5. Remove the clay bed or walls, being careful to keep the first jacket piece in situ.

6. Apply release agent to the jacket edge, and to the second piece of the jacket built against the first.

7. The mould is now complete, the outside jacket in two pieces firmly holding all the smaller pieces within.

MOULDS MADE USING CLAY WALLS

When moulding very large or immovable patterns it may not be possible to create clay beds to mask off the sections to be moulded. This method uses walls to mask off the sections of the pattern as the mould pieces are created.

When using this method, it is usually only suitable to use the laminating method of applying the plaster, as the walls may be built vertically onto the surface of the pattern. Plaster of the right thixotropic consistency (thick enough not to run off a vertical surface) can be built up slowly in layers. The first layer may run a little, but will deposit enough thickness to allow a quite rapid build-up with subsequent coats (see Chapter 2 for mouldmaking principles).

1. Usually clay is used as the wall building material as it is flexible and will adhere to most materials temporarily. However, theoretically any material could be used (wood, card and so on), as long as it can be fixed tightly to the pattern with no gaps for plaster to seep through, and can be removed cleanly without any permanent damage to the pattern.

2. Rigid materials will need to be cut approximately to the contours of the division lines on the pattern, and will usually need to be sealed with clay at the final point of contact to the pattern. Bear in mind that this method will usually be applied to patterns that may be too large to move, so that walls may have to be built onto a vertical surface. They will therefore need to be fixed firmly in place while making the mould pieces. The width and thickness of the walls will largely depend on the size of the original, but should be of a width to incorporate a large flange (at least 50mm) around each mould piece, to provide a wide surface area and fixing point between mould pieces.

3. As when creating a clay bed, keep the edges of contact perpendicular to the pattern on the high point of any under-cutting. When creating the mould pieces this will ensure that the mould edges are not feathered, and that the pieces are slightly plug shaped so that they come apart easily.

4. Shallow depressions can be made at intervals along the clay wall to provide nipple and cup registration between mould pieces.

5. If moulding a standing pattern that has a base, the walls will need to be taken down to the modelling board for each section. Build the pieces of the mould against the modelling board at these points, creating the mouth of the mould through which it will be filled. If moulding a pattern completely in the round, attach a stalk of plasticine to the pattern. When the mould is complete and the stalk removed it will leave a hole through which the casting material can be poured. Stalks should be positioned on a section of the pattern where the subsequent casting can be easily reworked, as there will be a loss of detail at this point. Stalk diameter will depend on the size of the pattern and what casting material will be poured into the mould.

6. Apply appropriate release agents to walls and pattern.

7. With a first pattern section walled, make up the mould following the same method as for laminated 'One-piece reusable moulds', steps 2–6. The mould should be built out at least 50mm onto the walls to provide a wide flange around the mould piece edges. Allow the first mould piece to set thoroughly.

8. Remove the first set of walls, ensuring the first mould piece stays in situ on the pattern.

9. Attach a new set of walls to define the next section of the pattern to be moulded. The next mould piece to be made will connect with the first mould piece. A section of the walls defining the new section will be a section of the flange from the first mould piece. The walls should be built up tight to the first mould piece at this point. If a mould is to be made in just two pieces the flange of the first piece provides the wall up against which the second piece will be created (see 'multi-piece mould' photo sequence for illustrated principles).

10. Apply release agents to the new walls and the exposed flange.

11. Follow the method as for step 7.

12. If more mould pieces are to be made, repeat steps 8–9 until all mould pieces are created, each piece connecting to another to complete the mould.

13. If the mould is in the round do not forget to create a pouring hole in one of the mould pieces (see step 5).

14. For opening, trimming and cleaning, see above.

Press moulded cast from log.

CHAPTER 5

PRESS MOULDS

Clay press moulding is probably the most fundamental method of mouldmaking, in that it takes a negative from a positive by the simplest of means, that of using wet clay to produce a mould from a pattern, and then casting directly into that clay. Clay as a moulding material has similar properties to synthetic moulding materials (for example, silicone, Gelflex), in that it can to a certain extent be bent around undercutting and then retain its pattern shape.

Generally, plaster is used to create the casting from the clay mould because it can be produced quickly before the clay mould starts to dry out and no release agents are required. It is usually a one-off mould that is created, meaning that the mould is destroyed during removal following the first casting.

Any sort of clay can be used, but bear in mind that red clays may colour the casting (which can be interesting!). The consistency of the clay used, however, is important – too soft and the clay will break up and smear when released from the pattern, but too hard and a good impression will not be achieved.

This is not a mouldmaking process of complete accuracy when compared to the more high-tech moulding processes, and there are certain restrictions in terms of the amount of undercutting and detail that it will accommodate. However, these imperfections can produce unique and interesting castings, and quite a considerable amount of detail can still be achieved with such a basic method of mouldmaking.

There are three methods: bed, block and case. Which method to use depends upon the size of the pattern and the amount of undercutting:

- **Clay Bed** A clay bed into which small patterns in relief or with little undercutting are pressed.
- **Block** A small block of clay is pressed onto small patterns or small areas of a pattern in relief or with little undercutting.
- **Case** Slabs of clay are pressed onto patterns and supported with a rigid case. Suitable for larger and more undercut patterns.

Clay Bed Moulds

This method is suitable for small patterns with minimal undercutting. Patterns that are in the round or with deeper undercutting can only be partially moulded, because when they are removed from the clay the clay mould will be damaged.

1. Prepare a clay bed of the right consistency, at least 50 per cent thicker than the pattern to be pressed. The size of the bed will depend on whether the pattern is to be cast onto a background plaster slab as a relief or as a separate object (*see* diagram 1). As a relief, the clay bed should be the same size as the background plaster slab required; as a separate object the bed should be approximately 50mm larger than the pattern all round.
2. Dust the pattern generously with talcum powder to prevent it sticking to the clay bed, and brush off excess so that just a light dusting remains (excess talc will clog detail and compromise moulding detail). It may be necessary to experiment with other release agents in order to achieve the best results (*see* Chapter 3).
3. Press the pattern firmly down into the clay bed, avoiding any lateral movement. Patterns that are in the round should only be pushed to the high point of any undercutting (*see* diagram 1).
4. Gently lift out the pattern from the clay bed, again avoiding any lateral movement that may change the first impression in the bed.

Diagram 1. Clay press mould from a log

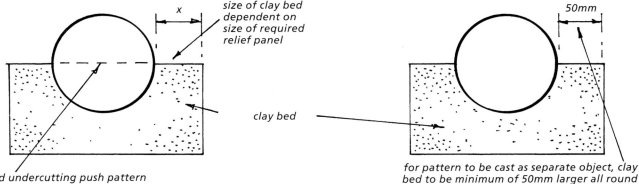

size of clay bed dependent on size of required relief panel

to avoid undercutting push pattern up to half-way into clay bed

clay bed

for pattern to be cast as separate object, clay bed to be minimum of 50mm larger all round

5. If a relief on a background is to be cast, create a clay retaining wall around the clay bed to contain the plaster when poured. The height of the wall will determine how thick the relief panel can be poured (*see* diagram 2)
6. The mould is now complete and plaster can be poured in order to create the cast (*see* Chapter 2 and relevant casting chapter).

Removing the Cast

1. Having allowed the plaster to harden, remove the cast from the mould. If the original pattern was not too detailed or undercut it may be possible simply to lift the cast out of the mould, a little suction being the only hindrance.
2. If there is a lot of suction or detail and undercutting preventing the cast coming out cleanly without damage, it may be necessary to remove the mould from the cast. To do this, invert the clay mould and cast, and, holding them together, gently pick away the clay mould from the cast.

3. Once free from the mould, trim and clean the cast (*see* Chapter 10).

Block Moulds

This method is suitable for moulding small sections of a large pattern that could not be practically pressed into a clay bed or cannot be moved, for example sections of architectural detail or natural detail such as a tree trunk. This method should not be used on patterns with deep undercutting. A clay block is pressed onto the pattern to capture the detail required, and is then removed to cast into. Only small moulds should be attempted, otherwise the mould may go out of shape when setting up for casting.

1. Dust the surface of pattern with talc; blow off the excess.
2. Prepare a block of clay of the right consistency. The block should be at least 50 per cent thicker than the depth of

Diagram 2

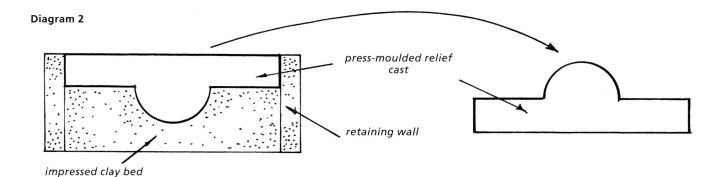

press-moulded relief cast

retaining wall

impressed clay bed

detail on the pattern to be moulded. The length and width of the block will depend on what can be removed from the pattern while safely retaining its shape when laying it back down. This may largely be a case of trial and error, but a block of clay over 200mm long will be difficult to handle. Whatever size is used, it should be a block that when pressed onto the pattern will retain a flat base, to lay back down flat onto a modelling board when removed. If, when pressing the block onto the pattern it bends, it may be possible to cut the base flat with a clay wire if there is enough thickness to the block (*see* diagram 3).

3. Press the block firmly onto the surface of the pattern, ensuring an even pressure is obtained to pick up all the detail.
4. Once pressed, carefully remove the clay block. Use the flexibility of the clay to bend it carefully around any undercutting, but remember that it will need to be bent back into shape when laid down again (this is where the size can be restrictive).
5. Lay the clay block straight back down on a modelling board, ensuring the back is flat to the board.
6. A clay retaining wall the thickness of the cast required can now be built around the clay block and the plaster cast can either be poured or laminated into the mould (*see* Chapter 10).

Removing the Cast

This is as described above in 'Clay Bed Moulds'.

Case Moulds

This method of press moulding is suitable for larger and more undercut patterns where it would be impractical and inaccurate to use the clay bed or block methods. It involves creating the mould with slabs of clay pressed onto the surface that are then supported in a rigid plaster case once removed from the pattern.

The mould is created gradually in slabs 30–50mm thick laid up side by side. Each slab, although laid up tight to the next, remains separate, enabling the slabs to be removed separately when the mould is complete. As the clay slabs are flexible, they can be bent around undercutting during removal from the pattern and then carefully laid into the plaster case to regain the original pattern shape. In this way, quite large or undercut patterns can be moulded. One-piece or multi-piece moulds can be made using this method.

One-Piece Case Moulds

1. Prepare the pattern with a dusting of talcum powder to prevent the clay sticking; blow off the excess.
2. Cut slabs of clay from a block of clay using a wire or clay harp. The consistency of the clay should be a little firmer than for previous two methods, otherwise the slabs will break when being removed from the pattern. The length and width of the slabs will depend largely on the size and complexity of the pattern to be moulded. Less undercut moulds may be made up of larger slabs; more undercut patterns will need smaller slabs. Bear in mind that all the slabs will eventually have to be lifted intact off the pattern. Slabs larger than 100mm wide by 500mm long will be hard to handle. Thickness should be between 30–50mm depending on undercutting; the slabs need to be thick enough not to thin out too much when pressed into the detail of the pattern.
3. Press the first slab firmly onto the surface of the pattern, making sure to press into any undercutting.

Diagram 3

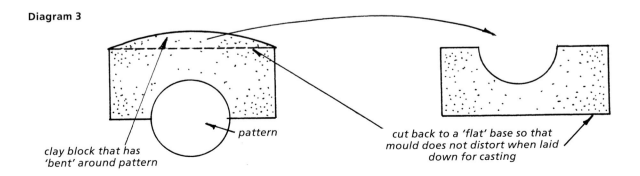

clay block that has 'bent' around pattern

pattern

cut back to a 'flat' base so that mould does not distort when laid down for casting

4. When laying on the next slab, butt it up as tightly as possible to the first without merging the clay together, so that each slab can be removed separately later. In this way, completely cover the pattern in a blanket of individual clay slabs. It is important to butt the slabs together tightly, although there will still be some fine raised lines on the casting caused by the fine gaps between the slabs. These can be easily rubbed off, but butting the slabs together as tightly as possible will reduce the problem.

5. Once the whole surface of the pattern has been covered, secure a retaining wall around the perimeter of the slab blanket to retain the edges of the plaster case as it is made.

6. Place a cling film membrane on the clay blanket. This will enable the plaster case to be removed easily later, leaving the blanket in situ.

7. Build up the plaster case on the clay blanket using the laminating method (*see* Chapter 4).

8. Once the plaster case has set fully and hardened, the mould can be removed from the pattern ready for casting (*see* Chapter 2 and relevant casting chapter).

MOULD REMOVAL AND REASSEMBLY

1. Carefully remove the case from the slab blanket. Ideally, the case should come away separately from the clay (the cling film will aid this process), enabling the clay slabs to be removed separately and then laid back into the case.

2. In this way, the slabs may be bent around any undercutting while removing from the pattern and then reformed when laid back in the case. If the pattern is not too undercut, the case and slabs may be removed together without distortion of the clay.

3. Remove the slabs from the pattern methodically one at a time and carefully lay back into the case for support. As they are laid back into the case ensure they go back in the same sequence they were put onto the pattern, and that they are butted up tightly to one another as they were made.

4. Once all the slabs have been laid back into the plaster case the mould is ready to be cast in. The casting should be made immediately before the clay dries out. It is usual to make a plaster casting from a press mould, but theoretically any casting material could be used as long as the appropriate release agents are applied (*see* Chapter 3).

N.B. It is possible to create a press mould by laying up one large blanket of clay that can be cut into more manageable strips to

lay back into the case. A blanket can either be built up in chunks of clay pressed tightly together onto the pattern or made in one large slab. A case is then constructed as before. When the case has been removed, the clay blanket is cut through to the surface of the pattern in strips and laid back into the case for casting.

REMOVAL OF THE CAST FROM THE MOULD

1. Removal of the cast from the mould should be treated in the same way as for any flexible mould, that is, the rigid case should be removed first to allow the full flexibility of the moulding material to be utilized.

2. Carefully remove the plaster case, and then gently remove the slab blanket from the casting. Press moulds are generally one-off moulds. However, it may be possible to reuse the mould if the blanket is not too damaged during removal, and can be reassembled back into the case.

3. There may be some fine raised lines on the surface of the casting, a result of the clay chunks or slabs not being butted together tightly enough, but these can be easily removed with tools or abrasives.

Multi-Piece Case Moulds

Patterns that are deeply undercut or in the round will require a mould in more than one piece to allow for removal of the mould from the pattern and removal of subsequent castings. This is carried out as for other moulds made in pieces.

1. Determine the mould division lines on the pattern, either by marking on the surface or by committing to memory.

2. Secure the pattern to a modelling board with the first section of the mould to be created uppermost.

3. Work within the lines determined for each piece of the mould. Slabs of clay are laid and pressed into the detail of the first section of the pattern to be moulded, a retaining clay wall is built and a plaster case is created over the slabs (*see* 'Laminate Moulds' p.41, steps 2–6).

4. When making the case for multi-piece moulds it is important to create a flange all the way round to secure the mould pieces together while casting. The retaining wall around the slabs needs to be of a sufficient height to incorporate this flange. Make shallow depressions at intervals along the retaining wall to form nipple and cup registration points between mould pieces.

5. Once the first mould piece is complete, rotate the pattern to expose the next section of the pattern to be moulded. Take care to keep the first piece of the mould in place on the pattern as you do this.

6. Apply release agent to the exposed surface of the case flange.

7. Repeat steps 1–6 for each section of the pattern to be moulded, methodically creating each piece of the mould against the next until the whole surface of the pattern has been moulded.

8. Consideration should be given to access into the mould for casting – it may be necessary to create a pouring hole in one piece of the mould. If the pattern has a footprint, the mould mouth should be created around this (*see* Chapter 2 for mouldmaking principles).

MOULD REMOVAL AND REASSEMBLY

Before opening the mould it is important to consider how the pieces of the mould will be held together while casting. With the smaller moulds it may be adequate to use large rubber bands or cut-up lengths of bicycle inner tube, tied and made into bands of appropriate lengths. For the larger moulds the mould pieces can be bolted together – this is where a wide flange is vital. A wide flange will provide a large surface area of registration between mould pieces, but also plenty of space to drill holes to take a nut and bolt. If using nuts and bolts to secure mould pieces, all the holes to take them should be drilled before opening the mould to ensure accurate reassembly. Any hand or electric drill with a standard HSS twist drill bit can be used to do this. Bolts with wing nuts should be used for ease of reassembly, and a large washer on each side of the bolt (all should be galvanized so as not to rust).

The mould can now be opened.

1. Cut all flanges back by a few millimetres to expose the seam between them and provide a strong, tidy edge. This can be done with a surform. The excess plaster should be taken back until a line can be seen between the case pieces. All the seams exposed the case pieces can be removed.

2. Start by inserting a thin, strong-bladed knife in-between all the case pieces. Initially, this is just to part the adjoining flange seams of each piece, not to remove the pieces from the pattern.

3. Once the flanges have been parted the seams can be opened up more by methodically working around the seams. Open the seams equally and gradually all the way along, rather than in one place. As the seams start to open it may be necessary to use a chisel, particularly on larger moulds. Larger moulds may benefit from a series of wooden or plastic wedges being gradually driven into the seam in sequence until the case pieces separate. Once one case piece has been removed the others will generally come away easily.

4. Once the case pieces have been removed, the slabs can be removed and laid back into them.

5. Peel the clay slabs carefully away from the pattern and immediately lay them back into the case. It is important to lay the clay slabs back into the case exactly where they came from so that the mould surface is the same as when it was pressed onto the pattern.

6. Now that the mould has been removed from the pattern, a casting should be taken before the clay dries out.

MULTI-PIECE CASE MOULD

The original pattern.

Dusting with talcum powder.

MULTI-PIECE CASE MOULD *continued*

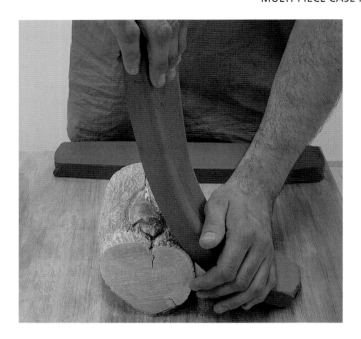

Laying on clay slabs.

Apply pressure on the clay slab to press into the pattern detail.

Creating a retaining clay wall around the area of pressed slabs. Lay clingfilm over clay slabs.

Application of the first plaster layer by brush.

Once the plaster case is complete, the mould and pattern are inverted to begin work on the second half of the mould.

The clay slabs are pressed to create the second half of the mould.

MULTI-PIECE CASE MOULD *continued*

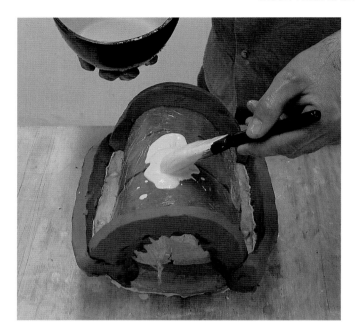

The retaining wall created, the release agent is applied to the exposed plaster of the first half and the second case part is begun.

The mould completed, the pieces are separated.

This half of the case has separated, leaving the slab still in situ.

Carefully remove the slab and lay it back into the plaster case.

The mould pieces are reassembled.

Because the mould base is not level the modelling board has been raised on one side. The mould pieces are secured with lengths of bicycle inner tube, then the plaster is poured.

The mould is separated, exposing the set plaster cast.

The clay slabs are peeled away.

One-piece silicone mould.

CHAPTER 6

SILICONE MOULDS

Moulding silicone comes as a pourable viscous liquid that will set to rubber with the addition of a catalyst. There are different silicone rubbers available that will set to varying firmnesses, and so can be used for a variety of moulding and casting applications (*see* Chapter 2).

Being a flexible medium, silicone moulds can be made in fewer pieces, the flexibility accommodating a certain amount of undercutting. Silicone is a very durable material with a long mould life, and so is suitable for long cast runs. Silicone is a room-temperature vulcanizing rubber, which means that it will set at room temperature with the addition of a catalyst and is not exothermic (will not generate heat). This can be advantageous when creating moulds from patterns that may be damaged if a hot-melt vinyl rubber were used. Silicone is a self-releasing moulding compound that contains release agents integrally and therefore does not need to be prepared with release agents prior to casting.

Silicone moulds can be made in blocks, by pouring within a retaining wall, or built up as a skin supported by a rigid case of plaster or fibreglass.

- Many silicones can be mixed to give intermediate flexibilities of set material, although batch testing should be carried out.
- Release agents are not normally required when casting from silicone moulds (*see* Chapter 3). However, prolonged use of certain casting materials (polyester/epoxy resins) will deposit a residue of material that may impede the releasing properties of the silicone. Mould life can be prolonged in this case by heating the mould in a cool (no more than 100°C), well-ventilated oven for two hours.
- Release agents may be required on certain pattern materials; for example, silicone will adhere to silicone, so when creating multi-piece moulds a release agent will need to be applied between pieces.
- Silicone will cure at room temperature; low temperature and humidity will retard curing, while high temperature and humidity will accelerate it.
- Silicone is a relatively safe material to use provided common-sense handling and product safety guidelines are followed.

The Material and Principal Considerations

- Silicone rubber should be stored in sealed, air-tight containers and has a shelf life of approximately six months at room temperature; refrigeration and freezing will increase shelf life considerably.
- The catalyst should not be subjected to long-term exposure to air, and should be stored in an air-tight container when not in use.

Measuring and Mixing

Silicone rubbers require the addition of a catalyst to turn them from a viscous liquid into a solid material. The amount and type of catalyst needed will depend on the type of mould being produced. Different catalysts added at different ratios to silicone will produce faster or slower setting times. If silicone is to be poured, a faster setting time may be advantageous. If it is to be built up gradually as a skin, then slower setting times will be necessary. Silicone that is built up in this way will also require the addition of a thixotropic additive (thixo), which will prevent the material running off the pattern as it is applied.

The addition of a catalyst and thixo to batches of silicone needs to be accurate. Too little catalyst, and the mould will not cure, which can be hard to rectify and an expensive mistake; too much thixo, on the other hand, will produce a mix that is too stiff to penetrate pattern detail.

Different manufacturers of moulding silicone will provide different methods of measuring catalyst and thixo additions (for example, 'drops of catalyst to weight of silicone', 'weight of catalyst to weight of silicone', 'percentage of thixo to weight of silicone'), but whichever method of measurement is employed accuracy is vital. The chart provided below displays weight of silicone to weight of catalyst, which is probably the most accurate method as long as accurate scales are used. Usable batches of silicone will be between 125–250g with additions of between 12–50g of catalyst depending on the curing time required, so scales in increments of 1g are essential. Good-quality balance or digital scales are ideal. The curing time of batches will depend on which ratio of silicone to catalyst is used. Generally speaking, a 20 : 1 ratio is suitable for building up skin moulds, and a 10 : 1 ratio for pouring block moulds.

The chart shows additions of thixo as a percentage expressed in gram weight (2 per cent being equal to 10g of thixo). Additions of between 0.5–2 per cent of thixo can be added to give batches of silicone a 'brushable' consistency that will pick up pattern detail and can be built up to approximately 5mm thickness without running off. Additions of 5 per cent silicone oil will reduce viscosity and increase working and curing times if necessary.

Note that the chart provided is a general guideline that will be effective for most condensation silicone rubbers, but different manufacturers' specifications may vary. Test batches should therefore be carried out before application to a project. (Note specific gravity conversion examples to convert millilitres to grams.)

Weight of Silicone to Weight of Catalyst

Silicone Rubber-Catalyst Ratios
20 : 1 Ratio
Silicone : Catalyst
500g : 25g
10 : 1 Ratio
Silicone : Catalyst
500g : 50g

Thixotropic Additive Addition at 2 per cent Rate
Silicone : Thixo
500g : 10g

Specific Gravity (SG) Conversion Equation
1ml = SG number in grams
Example : Catalyst with an SG of 0.96
5ml of catalyst = (0.96×5) grams = 4.8g
Example : Silicone with an SG of 1.24
100g of silicone = $(100 \div 1.24)$ ml = 80ml

Tools and Equipment

The following items are needed when using silicone:

- scales, either good-quality balance or digital, calibrated in 1g increments for measuring silicone and additive;
- paper cups or plastic containers for dispensing batches of silicone; plastic containers can be reused once leftover silicone has been allowed to set, then peeled out of the container;
- plastic or wooden spatulas for mixing batches;
- 12–25mm good-quality decorating paint brushes for the initial application of silicone when making skin moulds;
- metal or plastic palette knifes for secondary application of silicone when making skin moulds;
- craft knife or scalpel for trimming set silicone.

Block Moulds
One-Piece Moulds

One-piece are the simplest silicone moulds to produce. The mould consists of a wall created around the pattern; the silicone rubber is then poured around the pattern and allowed to set, creating a mould of a solid block of silicone rubber.

1. Secure the pattern to the workbench or modelling board, if necessary gluing it down (light objects will float to the surface).
2. Build walls around the pattern, with a gap of approximately 8mm around and 8mm higher than the highest point on the pattern. Follow the contours of the pattern to conserve

silicone. The walls of the mould can be made from a number of different materials, including clay/plasticine, MDF (or other wood or wood-type materials), Lego bricks, brass shim, cardboard/foam board and so on. Thought should be given to the rigidity of the material in relation to the size of the job, for example a larger pour will require more rigid walls in order to contain the volume of liquid rubber.

3. Thought should also be given to the releasing properties of the wall material; clay, for instance, will need no release agent, but plaster will need shellac and a non-silicone-based wax (see Chapter 3).
4. Ensure that there are no gaps between the walls, as silicone will seep out of even the smallest gap.
5. When pouring the rubber within the walls, it should be poured steadily and continuously from one point, preferably in a corner away from the pattern. This will push ahead the air in the mould and allow the rubber to enter any undercuts. As the walls fill, any air bubbles should rise to the surface and out of the mould.
6. When the mould is approximately half full stop pouring and allow the rubber to level.
7. Finally, continue to pour until the pattern is completely covered by at least 8mm of rubber.

OPENING AND CLEANING

1. Remove the clay walls before inverting the mould.
2. Moulds may stick slightly to the work surface, but can easily be peeled off. Depending on the detail on the pattern the mould may either come away, leaving the pattern in position on the work surface, or may come up with the mould and have to be removed once away from the work surface. If there is difficulty removing the pattern from the mould, cut it slightly to allow removal, which will also aid removal of subsequent castings (see diagram 1, Chapter 7).
3. Once empty, clean the mould with warm water (do not use detergents as they will shorten the release agent's life). It is now ready to use.

SPLITTING A ONE-PIECE MOULD

On some occasions, a one-piece mould can be split after it has set, in order to facilitate the removal of a cast. This would be necessary if you had a tall, narrow mould (for example, a candle mould), from which it may be difficult to remove the cast. In this case, a cut halfway down one side (or two sides if nec-

essary) can be made into the set rubber (using a scalpel or razor blade) in order to open the mould mouth wider to facilitate the removal of casts from the mould. It is a good idea to wave the line to provide registration. The mould can be held together at the cut point by tape or rubber bands, taking care not to over-tighten and distort the mould (see diagram 1, Chapter 7).

Two-Piece Moulds

Because of the flexibility of modern silicone, a lot of mouldings can be made in one piece. However, there may be a number of reasons why a mould in two or more pieces may be required, for example the size and depth of the undercutting. When making a two-piece mould, it is necessary to mask parts of the pattern while making each piece. This is achieved by embedding the pattern in a clay bed, leaving the part of the pattern to be moulded exposed and the rest of the pattern masked within the clay bed (see Chapter 2 for mouldmaking principles).

1. Deciding the division line of the mould will involve looking at the pattern and picking out a line around it at the high points.
2. Place the pattern on the bench or modelling board and build a clay bed up to the line. The surface of the clay bed must be built up tight to the division line to prevent silicone seeping through to the masked side. Use cling film to protect the side of the pattern that is to be embedded.
3. Care should be taken to ensure that the bed is perpendicular to the pattern and not feathered to a thin line.
4. Build out the bed between 8–20mm (depending on the size of the pattern) from the division line, following the shape to conserve rubber.
5. To ensure accurate registration between mould parts and to prevent cast leakage, registration points and pinch lines should now be created in the bed. These take the form of shallow depressions and raised lines on the surface of the clay bed.
6. Consideration should be given to access into the mould for casting. If the pattern has a base, secure a board tightly to it before embedding. The board will mask the base while the mould is created, providing the mould mouth when complete (see diagram 3, Chapter 7).
7. If there is no base, a pouring point into the mould will need to be created. This is done by connecting a cone of clay to

65

an inconspicuous part of the pattern; once removed from the completed mould, this will provide a pouring hole. At this stage, only half of the pouring hole needs to be created, the second half being created in the second piece of the mould (*see* diagrams 2a and b, Chapter 7).

8. Create walls around the bed (*see* 'One-Piece Moulds', steps 2–4).
9. Pour in the silicone (*see* 'One-Piece Moulds', steps 5–7).
10. Once the rubber has set, the first piece of the mould is complete.
11. To make the second piece, taking care to keep the pattern within the first piece of the mould, invert the clay bed and rubber.
12. Carefully remove the clay bed and cling film, exposing the masked side of the pattern sitting within the first part of the rubber mould.
13. If a clay cone has been attached for the pouring hole, create the second part of it to match the first part.
14. Apply a spray silicone release agent to the exposed silicone of the first piece to prevent the second piece sticking.
15. Create walls around the first silicone piece, 8mm higher than the highest point of the exposed pattern and pour the silicone as before.

OPENING AND CLEANING

1. Remove the clay walls and base board if necessary.
2. Carefully peel the mould pieces apart.
3. Remove the clay cone if necessary.
4. Wash the mould in warm water.

Block Moulds with a Rigid Case

As explained in the previous section, block moulds are created by building a wall around the pattern model and pouring rubber over and around it. Leaving a gap of 8mm around and above the model is usually sufficient to provide a mould with enough rigidity to support itself and not distort while casting.

If the model is particularly large, a thicker mould may be required in order to support itself while casting. This can be prohibitive in terms of cost, but also in terms of handling and setting up the mould without distortion prior to casting. The solution is to produce a rigid case around the rubber that will support it during cast production.

1. The block mould is produced in the same way as in the previous section by creating walls around the model and pouring on silicone rubber. Once the rubber has fully cured, the case can be produced.
2. Remove the walls containing the rubber mould and trim away any strands or thin pieces of silicone using a scalpel.
3. It may be necessary to cut registration keys and add registration blocks of pre-set silicone at this point to register the rubber into the case (particularly around the base, which when inverted will become the opening to the mould) (see 'Cutting Registration Keys' steps 2 and 3, p.69).
4. Create a new set of walls, approximately 8mm out from the edge of the set rubber mould and at a height of approximately 8mm higher than the highest point of the rubber mould.
5. Application of a release agent to the inside of the walls will need to be considered depending on the material from which they are constructed (*see* Chapter 3).
6. Mix some plaster and pour it over the rubber mould to the top of the walls. Either knock the mould gently or agitate the plaster surface to bring up any air bubbles (*see* Chapter 10). Alternatively, create a laminated case (*see* 'Laminate Moulds' steps 2–6, p.41).

OPENING AND CLEANING

1. When the plaster has set fully, remove the retaining walls.
2. Invert the mould and carefully remove the pattern. If it is difficult to invert the whole mould or to remove the pattern, the case should be removed from the rubber first, allowing more flexibility for removal of the pattern.
3. It may be necessary to make minimal cuts into the rubber to ease removal of the pattern and subsequent castings (*see* the previous section on splitting a one-piece mould).
4. Once the pattern has been removed, both the plaster and rubber mould parts can be trimmed, cleaned and dried ready for cast production.

Skin/Case Moulds

As the name suggests, these are moulds made up of a skin of silicone rubber supported within a rigid case. Because the rubber is being used in smaller amounts as a skin, rather than a block, these moulds are much more economical. As a skin, the silicone is much more flexible, enabling patterns with much deeper undercutting to be moulded. However, because of the increased flexibility of the skin, a rigid case (plaster or fibreglass) needs to be used in order to support the mould while casting. Other advantages of this system are the light weight and the accuracy of registration between multi-piece moulds.

Silicone rubber is a viscous liquid requiring a catalyst to make it set. With the addition of a thixotropic additive, the liquid becomes a brushable paste. The paste can then be brushed directly onto the pattern and built up using a palette knife. Additions of different percentages of thixo will produce different consistencies of silicone rubber. A 10 per cent addition by weight will allow the rubber initially to be applied by brush, achieving full penetration of pattern detail. Switching to a palette knife, the mixture can then be built up to full thickness (5–10mm) depending on mould size without it slumping or running off the pattern. An important property of mouldmaking silicone rubbers is their high tear and elongation strength (the degree to which the material can be stretched before breakage). Most silicone rubbers will have more than enough tear strength at this thickness to cope with quite deep undercutting. Silicone should be applied thickly over pattern undercutting to avoid cases undercutting.

One-Piece Moulds

These moulds are used for patterns that are usually either in low relief or have minimal undercutting, allowing access for cast removal during cast production. The area of the pattern that is in contact with the modelling board is known as the footprint or foot; this will become the opening into the mould. When deciding on a one-piece mould, an important consideration is whether the footprint is large enough to allow access in and out of the mould while casting (*see* Chapter 2 for mouldmaking principles).

1. Secure the pattern to a clean modelling board as previously described. Care should be taken to fill any gaps between the foot of the pattern and the modelling board so as not to allow any silicone to seep under the pattern.
2. Consideration should be given as to whether the pattern will require a release agent (*see* Chapter 3 for principles).

LAYING UP THE SILICONE

Covering Layer (First Coat)

1. Batches of 250–500g of silicone are easily workable amounts and should be mixed as per the mixing instructions above. Depending on the detail of the pattern, batches of different consistencies can be made with the addition of between 0.5–2 per cent thixo.
2. If the pattern is very detailed or includes deep undercuts, a batch containing just silicone and catalyst should be applied

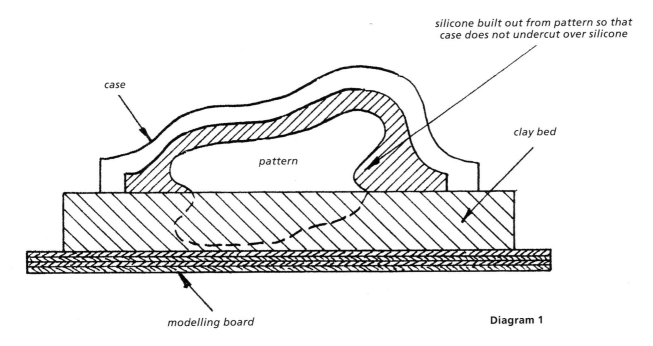

silicone built out from pattern so that case does not undercut over silicone

case

clay bed

pattern

modelling board

Diagram 1

first. This will produce a fairly runny mix that will easily flow into undercuts and cover any fine detail. Apply this initial layer with a soft brush (a 12–25mm good-quality decorating brush is suitable). Brush strokes should be short and circular in motion, slowly and methodically working over the surface to ensure that air is not trapped during the application.

3. This first layer can be applied in as many batches as necessary to cover the entire surface of the model. Batches with thixo addition should be applied to a depth of approximately 2–3mm; batches without thixo should not be more than approximately 1mm thick, otherwise they will run off the pattern and pool around the sides. A little pooling around the sides is inevitable if using batches without thixo, so do not worry if the silicone covering the pattern is less than 1mm thick as long as it is covered.

4. Paint silicone over the entire surface of the model and continue from the edge of the pattern onto the modelling board to a width of between 4–6mm. This will create a registration flange for the registration keys to be cut into when the silicone has set.

5. If using batches without thixo it may be necessary to apply another layer to reinforce the first. The first layer should be allowed to set fully before application of the second, and the second allowed to set before application of reinforcing layers (see below). Batches with thixo additions do not need to be allowed to set before the application of reinforcing layers.

Reinforcing Layers (Secondary Coat) Once the covering coat or coats have been applied and allowed to set, the working surface of the mould is complete – that is, the detail of the model has been captured and is safe under the initial layers of silicone. However, these layers must be reinforced to provide strength and elasticity to the silicone skin. The total mould thickness should be built up to a minimum thickness of approximately 5–10mm, depending on mould size. This is quickly achieved with the introduction of the thixotropic additive.

1. Batches of between 250–500g should be mixed and applied with a palette knife.

2. Where there are deep undercuts on the pattern the silicone should be built up much thicker to enable release of the case from the set silicone without undercutting (see diagram 1).

3. A good tip for thickening parts where there is deep undercutting is to use hessian scrim or split fibreglass matting as a packing agent to bulk out the silicone. Small pieces of scrim or matting should be thoroughly saturated in silicone

(it may be necessary to use a mix of silicone without thixo for this) and firmly pushed into the undercut with a brush or tool, working them in well to expel any air bubbles. More silicone is now applied on top (a thixotropic mix can be used for this), so that the scrim is thoroughly sandwiched between layers of silicone. This method will fill large undercuts quickly and cut down on the amount of silicone needed, but should only be used once a 3–5mm layer of silicone has been applied.

4. When the full thickness of silicone has been achieved, thought should be given to registering the silicone rubber into the rigid case. On moulds where there are large, flat areas of silicone it will usually be necessary to add registration blocks at these points to anchor the silicone into the rigid case. Registration blocks are blocks of silicone, cut to an appropriate size for the mould, and stuck onto the surface of the unset silicone once the overall thickness has been reached. It is advisable to have some old moulds or pre-set slabs available that can be cut up for this purpose. The size and thickness of these blocks will depend on the mould size, but should be large enough to hold the rubber into the case when the finished empty mould is inverted. Blocks should be glued on with catalysed silicone on areas that are flat with little undercut. These will provide registration into the case and prevent the silicone skin from moving away from the inside surface of the case at these points.

5. Once the registration blocks are in place it may be necessary to smooth out any swirls and wisps on the surface of the silicone that were created during application. The surface should be relatively smooth to enable the silicone to be relocated easily into the case between castings. To prevent palette knives or other tools from sticking and bringing up swirls and wisps a lubricant should be used. Any oily product, for example a light cooking oil, will prevent tools from sticking. Clean any excess silicone off tools and dip them liberally into the oil. Immediately pass them over the silicone surface, smoothing any wisps and swirls. As soon as the tool starts to drag and bring up swirls again remove it from the surface, wipe off the excess silicone again and re-oil. Repeat as necessary until the whole surface is relatively smooth. Take care when smoothing around registration blocks not to move them. It is important only to do this oiling process after all the silicone has been applied as any oil between layers of silicone would prevent them from sticking to each other. One the oiling is completed, the silicone can be left to set fully (usually overnight) before the next stage.

CUTTING REGISTRATION KEYS

Once the silicone has fully set, registration keys can be cut, in order to prevent the silicone moving about in the case during cast production This is achieved by cutting a series of dovetailed keys into the flange of silicone surrounding the model.

1. Cut away the ragged edge of the flange to a good edge, where the silicone reaches its full overall thickness of between 5–10mm.
2. A series of dovetails can now be cut at equal intervals and of a size appropriate to the mould size. There should be enough to hold the silicone skin firmly into the case all the way round, but not so many that it would become a chore to relocate them when taking the skin in and out of the case during cast production. This done, the mould is ready for the case making.
3. Registration blocks (*see* previous section step 5) can be added at this stage but will need to be allowed time to set before case making.

CASE MAKING

Using silicone as a skin in this way is economical and utilizes the full flexible properties of the material. However, due to these properties, the silicone will need to be supported by a rigid case of plaster or fibreglass during cast production in order to retain the shape of the mould. Depending on size, either material can be used, although fibreglass is preferable in terms of weight and ease of handling.

1. The case will need to extend approximately 10mm out from the end of the dovetails, so a release agent will need to be considered where the case material will have contact with the modelling board.
2. Check that the dovetailing is firmly in contact with the modelling board, otherwise case material will seep under the silicone. A good tip if the silicone is lifting around the edges of the flange is to use a little petroleum jelly to stick the silicone down temporarily to the modelling board.
3. If a plaster case is to be used, a small retaining wall of clay 10mm out from the end of the dovetails will be needed to contain the first layers of plaster. Plaster/fibreglass can now be built up over the silicone (*see* Chapter 4 or 11 for specifications). Take care when applying the first coat of plaster or resin gel coat that the material thoroughly covers the silicone and penetrates between the dovetails to ensure a tight

fit between the case and silicone. Once the case has been made, the mould is complete and can be opened, trimmed and cleaned.

4. Please note that it is also possible to create silicone case moulds by pre-making a case and filling it with silicone rubber in the same way as for 'vinyl case moulds' (*see* 'Vinyl Mouldmaking' and substitute silicone rubber for vinyl rubber).

OPENING, TRIMMING AND CLEANING

1. Release the case from the modelling board using a palette knife and then carefully lift it away from the silicone skin. Be aware that on models with low relief and undercut this process will sometimes bring the silicone skin away with the case.
2. If the case comes away separately, the silicone can be peeled away from the pattern with ease using the full flexibility of the material. This is how the mould should be used during cast production, removing the case and then the silicone in two operations.
3. Once the mould has been removed from the pattern model, it should be examined. Use a scalpel to trim off any very thin strands or skins of silicone around the opening or within the mould, as they are likely to be pulled off during cast production.
4. Trim the case of any sharp or rough edges with tools that are appropriate to the material.
5. Thoroughly clean the case and silicone of any trimming debris and pattern material using warm water and a weak detergent if necessary (strong detergents should not be used on the silicone as they may affect the releasing properties of the rubber).
6. Once dried, the mould can be reassembled ready for use. Care should be taken to fit the silicone back into the case positively, ensuring that all the dovetails and any blocks are securely located within the case.

CASTING

It is advisable to test the mould with a plaster cast first. This is a quick and cheap way to see how the mould is working and whether it will need any adjustments to the case or silicone in the form of trims or cuts to ease removal of castings (*see* Chapter 2 and relevant casting chapter).

ONE-PIECE CASE MOULD

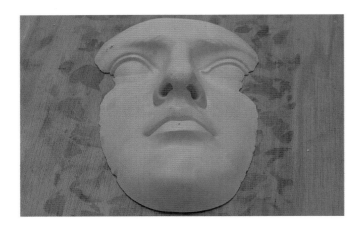

The original pattern.

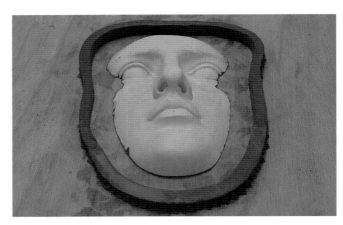

The clay retaining wall.

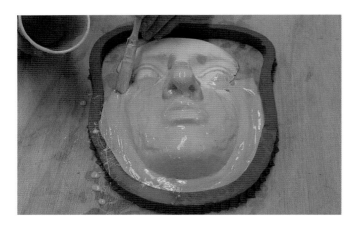

Applying the first layer of silicone.

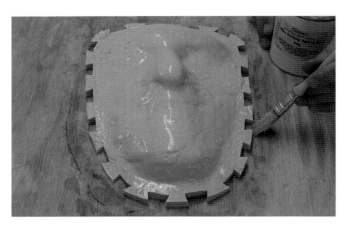

Dovetails are cut into the set silicone and release agent is applied to the modelling board.

A pigmented polyester resin gel coat is applied.

The gel coat is backed up with fibreglass and a pigmented laminating resin.

The set case is removed by slipping a thin blade between the board and case.

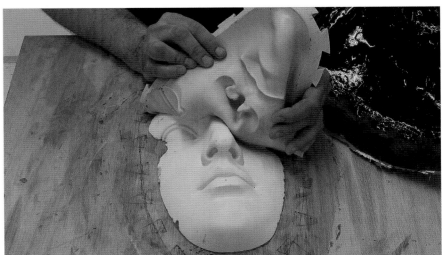

The silicone is peeled away from the pattern.

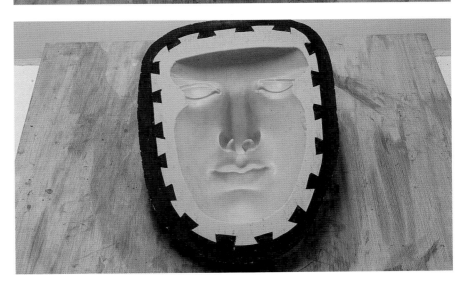

The finished mould, with the silicone back in the case.

Multi-Piece Moulds

Despite the flexibility of silicone, it is not always possible for the opening of a one-piece mould to stretch enough to allow access in and out of the mould during cast production, and patterns that are more complicated in form and have deeper undercutting will require a mould that is in two or more parts in order to release casts during cast production.

To make a mould in parts, one must first determine the lines of separation of those parts on the pattern's surface. Unlike plaster mould production, the division lines between mould pieces can be fewer and less precise due to the flexibility of the silicone, which will enable the mould to be stretched over some undercutting during cast production. But as with any mould that is in parts the division lines should generally follow the lines of high points on the pattern that divide deep undercuts. They should also be in areas where it will be easy to clean off any seam flashing that occurs during casting. Division lines can be temporarily scored onto the surface of clay patterns, pencilled onto others or just committed to memory. Once these division lines have been established, the first section of the mould to be made will need to be masked off with a clay bed or walls (see Chapter 2).

Clay Bed Moulds

FIRST PIECE

1. Secure the pattern to the modelling board with lumps of clay, with the first section to be cast facing upwards. Use cling film to protect the part of the pattern that is to be temporarily concealed within the clay bed.

2. If the pattern has a footprint, a baseboard of approximately 150mm larger all round than the base should be secured tightly to it. The mould will be built up onto the baseboard, with the footprint eventually forming the mouth through which the casting material will be filled. The baseboard will need to be relocated as each piece of the mould is made.

3. If the pattern is completely in the round and has no base, then a pouring hole into the mould will need to be created. Attach a truncated cone of clay at a point on the pattern with little detail. Build up the mould around this cone; when the cone is removed it will leave behind a pouring hole (see diagram 2a, Chapter 7).

4. Initially using large pieces of clay, build up the bed of clay from the board to just below the first section of division lines on the pattern.

5. Next, with smaller pieces of clay take the bed up to the lines following them around the form. The divisions on the pattern are unlikely to be in straight lines and will generally undulate around the form, therefore creating a bed of clay that will undulate at those points. This is perfectly acceptable, and in fact will form some registration between the mould pieces.

6. The clay bed should be built out from the pattern to a width of approximately 100mm, but can be larger depending on the size of the pattern. The width of the bed will need to incorporate a flange of silicone.

7. Once the clay bed has been built up to the division lines, smooth it off. The clay bed should be built tight to the pattern so that silicone cannot seep into other sections of the mould. The bed should also sit perpendicular to the pattern to avoid feathered edges between mould parts.

8. The bed complete, cut a half-round pinch line into the clay with a wire modelling tool to provide registration and a leak barrier between mould pieces. This should be created 20–30mm out from the division lines.

9. Apply release agents to the exposed part of the pattern (see Chapter 3).

10. The silicone can now be laid up, registration keys cut and a supporting case made as described above for the one-piece mould. Before making the case, nipple and cup registration points should be made in the clay bed using fingers or a round-ended tool handle.

11. A plaster or fibreglass case can now be made as for 'one-piece' mould.

SECOND PIECE

Once the first part of the mould is complete, the next section of the mould can be started.

1. By carefully inverting the clay bed complete, with the pattern sitting within the first mould piece created, the clay bed can be removed. It is important at this stage to keep the pattern within the first part of the mould so as not to disturb the registration continuity between the mould pieces.

2. With the mould inverted and stabilized on the modelling board, the clay bed can be removed by carefully picking the clay back to the surfaces of silicone and case material.

3. The exposed surfaces of the first mould piece will probably need cleaning up of clay and any cling film removed from the next section to be cast.

4. If creating a mould in more than two pieces the next set division lines should now be determined which will connect to

one section of the first mould part created. A new clay bed can now be created up to the new set of division lines and the one section of the first mould piece in the same way as previously described.

5. When creating the pinch line in the new clay bed it should start and finish at the exposed pinch lines created on the first mould part so they connect up.

6. Release agents should now be applied to the exposed next section of the pattern and the exposed surfaces of the first mould piece.

7. Create the next mould and case piece as previously described.

8. If a two-piece mould is being made the mould is complete, if the mould is in more pieces work methodically around the pattern until all the pieces connect and the pattern is covered.

OPENING, TRIMMING AND CLEANING

Before opening the mould it is important to consider how the pieces of the mould will be held together while casting. With the smaller moulds it may be adequate to use large rubber bands or cut-up lengths of bicycle inner tube, tied and made into bands of appropriate lengths. For the larger moulds the mould pieces can be bolted together – this is where a wide flange is vital. A wide flange will provide a large surface area of registration between mould pieces, but also plenty of space to drill holes to take a nut and bolt. If using nuts and bolts to secure mould pieces, all the holes to take them should be drilled before opening the mould to ensure accurate reassembly. Any hand or electric drill with a standard HSS twist drill bit can be used to do this. Bolts with wing nuts should be used for ease of reassembly, and a large washer on each side of the bolt (all should be galvanized so as not to rust).

The mould can now be opened, trimmed and cleaned.

1. All flanges should initially be cut back a few millimetres to expose the seam between them, and provide a strong, tidy edge. With plaster cases this can be done with a surform. Take the excess plaster back until a distinct line can be seen between the case pieces.

2. Once all the seams have been exposed, remove the case pieces. Fibreglass case seams can be cut back with a hacksaw blade. Start by inserting a thin, strong-bladed knife in-between all the case pieces. Initially this is just to part the adjoining flange seams of each piece, not to remove the pieces from the silicone.

3. Once the flanges have been parted the seams can be opened up more, by methodically working around the seams. Open the seams equally and gradually all the way along, rather than in one place. As the seams start to open it may be necessary to use a chisel, particularly on larger moulds. Larger moulds may benefit from a series of wooden or plastic wedges being gradually driven into the seam in sequence until the mould pieces separate. Once one case piece has been removed, the others will generally come away easily.

4. With all the case pieces removed the silicone sections of the mould can be removed from the pattern. Gently peel each one away from each other and the pattern.

5. Once the mould has been completely released from the pattern, trim the case pieces and sand them if necessary for ease of handling.

6. Wash the mould in warm water and dry it by hand.

7. Reassemble the mould securely to prevent it warping before it is used.

MULTI-PIECE CASE MOULD

The original pattern.

MULTI-PIECE CASE MOULD *continued*

A clay bed, pouring point and pinch line for registration
are created.

Silicone is applied, initially by brush and subsequent layers
by palette knife.

Smooth off the last layer of silicone with an oiled
palette knife.

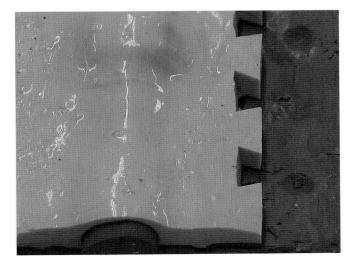

Cut dovetails into the set silicone and create registration cups in the clay bed.

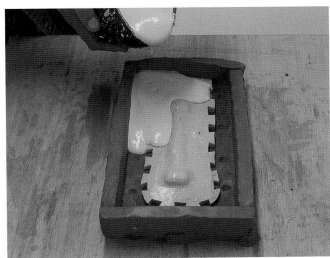

Pouring plaster within the retaining walls to create a case (an alternative case could be made by laminating plaster or fibreglass).

Once the plaster has set, invert the mould and remove the clay bed.

MULTI-PIECE CASE MOULD *continued*

Clean the moulded surface, then create the second half of the clay stalk for the pouring point.

After release agent has been applied to the first silicone mould piece, the second piece can be created.

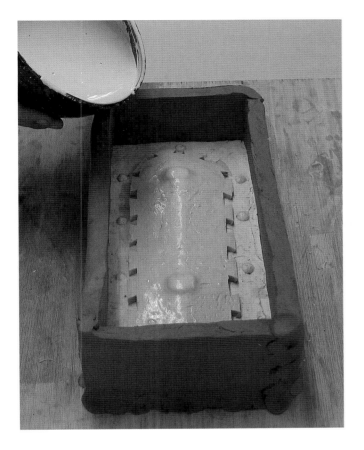

Once the dovetails have been cut, the retaining wall created and the release agent applied, the second piece of the case can be created.

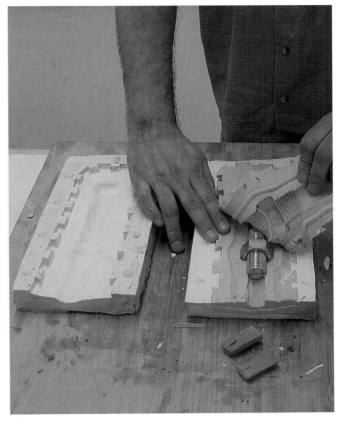

Separate the plaster case pieces and peel the silicone away from the pattern.

Wall Division Lines

When making multi-piece silicone moulds it is also possible to create pattern divisions for the mould pieces using walls instead of a clay bed. This can be advantageous, for instance, when moulding a large standing figure that would be difficult to lay down horizontally and would require very large amounts of clay to create a bed.

1. The same considerations need to be given to undercutting and pattern divisions as for clay bed construction and the mould pieces are created in the same way, one by one, each connecting to build up the mould.
2. Walls are used to mask the sections of the pattern as each piece of the mould is made. The wall material can be clay or any other material that can be temporarily but securely attached to the pattern as the pieces are constructed.
3. Walls should be wide enough to incorporate a flange of silicone and the case in the same way as for a clay bed.
4. It may be more advantageous to use the nipple and cup registration method between silicone pieces rather than pinch lines.
5. The walls should be attached as perpendicular as possible to the pattern surface.
6. The clay bed procedure above should be followed as for creating the mould and case pieces.
7. When moving onto the next section of the mould, remove the walls and rebuild them, defining the next section of the pattern to be moulded.
8. The mouldmaking procedures are then repeated until the mould is complete.

CHAPTER 7

VINYL MOULDS

Gelflex and Vinamold are reusable flexible mouldmaking materials that have the properties of rubber, but chemically are a *vinyl* plastic. Gelflex and Vinamold are effectively the same material; however, Gelflex has a less greasy surface and is a more pleasant material to use, and is the material referred to throughout this chapter. Either material can be used, but they should not be mixed.

Gelflex is a hot-melt moulding compound. It is heated from a solid until liquid, and poured over a pattern. It then resets to a solid to form the mould. It is available in several grades of flexibility, from soft to very hard; grades can be mixed to create intermediate grades of flexibility. It is a self-releasing compound, in that it contains release agents integrally.

The Material and Principal Considerations

Vinyl moulds can be made from a variety of pattern materials, but the materials may need different preparation before moulding. The easiest materials to mould from are wet (but not soaking) clay or metal patterns as these need no preparation. Porous materials such as dry plaster, stone or concrete patterns should be soaked in cold water for several hours prior to moulding to displace the air within the object. If a porous pattern is left dry, air will bubble out of the object and create pin-prick holes on the surface of the mould as the Gelflex sets. The problem of air within porous patterns can be restrictive as it may not be possible to soak certain patterns; for example, wood patterns may warp and crack, while other materials may dissolve. It may be possible to treat patterns that cannot be soaked in water with PS8 sealer, which is a light oil that can be painted onto the pattern coat by coat until the pattern is completely saturated (a light cooking oil can be used as an alternative). Soaking patterns in water or oil may affect their appearance permanently, which should be taken into consideration before proceeding.

Another restriction may be the high melt temperature of the Gelflex, which may crack or melt the pattern; for example, it would not be possible to use Gelflex on a pattern made of wax or very fine glass. However, larger glass or glazed ceramic patterns can be heated up slightly before moulding to reduce the heat shock.

The melting and pouring temperatures vary between grades, but standard grade Gelflex has a melting temperature of 160°C and a pouring temp of 150°C. The melting temperature of Gelflex can be restrictive to casting materials with a higher pouring or setting temperature (low-melt alloys for instance) and this should be taken into consideration.

It is important not to overheat Gelflex as it will burn the release agents out of the compound. Once a Gelflex mould has been used it can be remelted and used again to create another mould. It can remelted in this way seven to ten times depending on the casting material used in the mould – certain casting materials such as polyester resins are more destructive to the release agents within Gelflex.

There are a number of ways to melt Gelflex, but all must be controlled so as not to overheat the compound. Small quantities can be melted on a stove top using a saucepan within an air bath. An air bath is like a double boiler used for melting chocolate but without the water (boiling water would not be hot enough to melt the Gelflex). A cooking thermometer should be used to control the temperature carefully. Alternatively, a thermostatically controlled electric cooking pan can be used for small quantities. For larger amounts specially made Gelflex melters should be used. They are available for quantities from 5–50kg. Cut the Gelflex into pieces approximately 25–50mm (use larger melting vessels for larger pieces) and feed them gradually in to the melting vessel, stirring frequently as the required amount builds up. It is important to melt enough compound to create the mould in one pour (estimate for a little more compound than you need). If a slightly darker shade to the compound is visible during melting this is not a problem, but if it becomes a considerably darker shade throughout and fumes heavily, the compound is overheating and should be discarded. Gelflex will inevitably give off some light fumes and should only be melted in a well-ventilated area.

Once there is a sufficient quantity of Gelflex in the melting vessel the temperature should be dropped by 10° to the pouring temperature. The pouring temperature of Gelflex can be anything from 120–160°C, so great care should be taken. It is advisable to wear protective clothing and gloves while pouring.

Gelflex is a very versatile and economical method of mouldmaking and can be used for several moulding and casting applications.

N.B. Before embarking on any moulding method, Chapter 2 should be studied thoroughly.

Block Moulds

One-Piece Moulds

Probably the simplest of Gelflex moulds, block moulds should only be used for small patterns with little undercutting. As with other flexible moulding materials Gelflex will accommodate a certain amount of undercutting, the flexibility of the material enabling moulds to be pulled open around areas of undercut detail on a pattern. One-piece moulds are only possible with patterns that have a large enough foot to provide the opening into the mould (*see* Chapter 2 for mouldmaking principles).

1. All patterns except wet clay should be soaked or primed with PS8 sealer prior to moulding. Wet clay needs to be kept damp with a water atomizer during construction.
2. Set up the pattern securely on a modelling board or workbench. Light objects may float in the liquid Gelflex before it sets and may need to be temporarily secured, glued down if possible.

3. Construct a retaining wall around the pattern at a distance of approximately 50mm from the foot. With small patterns the wall can be made of slabs of clay cut with a wire harp. The wall should be 25–50mm thick and 25–50mm higher than the highest point on the pattern, depending on size, and should be pinched onto the surface of the modelling board or bench around the outside of the wall. For larger patterns, a more rigid wall may be required (MDF, ply or rigid modelling card, for example), which can be tied together with string or wire or screwed together and sealed to the work surface with clay to prevent leaks. Walls made from porous materials should be treated with a PS8/light cooking-oil sealer to prevent air seeping into the Gelflex while it sets (dense wall materials such as MDF require little or no sealing).
4. Once the retaining walls are securely in place, pour the Gelflex within the retaining walls. Avoid pouring directly onto the pattern and maintain a steady, even flow to allow air to be pushed out as the Gelflex rises within the walls. Enough Gelflex should be melted to allow the pouring in one operation without a break. If it is poured in several batches, when hot Gelflex meets cooler Gelflex from a previous pour, a small line, known as a chill line, will appear on the surface of the mould. Some chill lines are inevitable and will result in a very fine raised line on the surface of castings; these are easily rubbed off, but they can be kept to a minimum by pouring the Gelflex at the correct temperature and in one steady, continuous flow.
5. Allow the Gelflex to set fully before turning out the mould. Setting times will vary depending on the volume of Gelflex poured. Overnight cooling is preferable, but a minimum of three hours should be allowed.

OPENING AND CLEANING

1. Remove the clay walls before inverting the mould.
2. Moulds may stick slightly to the work surface, but can easily be peeled off. Depending on the detail on the pattern the mould may either come away, leaving the pattern in position on the work surface, or may come up with the mould and have to be removed once away from the work surface. If there is difficulty removing the pattern from the mould, cut it slightly to allow removal, which will also aid removal of subsequent castings (*see* diagram 1).
3. Once empty, clean the mould with warm water (do not use detergents as they will shorten the release agent's life). It is now ready to use.

Two-Piece Moulds

Patterns that have more detail and undercutting may require a mould made in two parts to allow the removal of the pattern and subsequent castings from the mould. Two-piece moulds are only suitable for smaller patterns as it would be difficult to register the two pieces together if they were too large. Two-piece moulds are made one piece at a time, a clay bed being used to mask one half of the pattern as the first mould piece is made.

Carefully consider the shape of the pattern, looking at where undercutting and detail occurs. A line defining two halves of the pattern should be determined along a contour with the least detail or difficult undercutting. Because of the flexible nature of Gelflex it is not necessary to create exactly equal halves, because the mould will bend around a certain amount of undercutting (*see* Chapter 2 for mouldmaking principles).

1. With the division lines determined, decide which piece of the mould is to be created first and secure the pattern temporarily to the modelling board or workbench with this section of the pattern facing up as horizontally as possible. It may be easier to lay the pattern down horizontally to do this.

2. Secure the pattern with clay, bricks or pieces of wood, depending on the size of the original, wedged around to keep it in place.

3. Build small pieces up to approximately 20mm below the lines defining the first piece of the mould. The bed should extend 150mm around the original. Initially, the bed can be created fairly roughly, gradually building up to the division lines with more care.

4. When approaching the division lines the bed should be smoothed off. Care should be taken to avoid curving the clay up at the final point of contact between the clay bed and the pattern, as this will create feathering. A clay tool should be used at this stage to create the bed perpendicular to the pattern. The bed may undulate following the lines of division.

5. Make shallow depressions at intervals around the pattern into the surface of the clay bed with a round-ended tool handle to provide registration between the mould pieces. Be aware when making registration points between pieces that if the pieces cannot be taken apart vertically from one another they will not come apart.

6. If the pattern is completely in the round, a pouring hole will need to be created. This is done using a truncated cone of clay attached directly to the pattern (*see* diagram 2). The

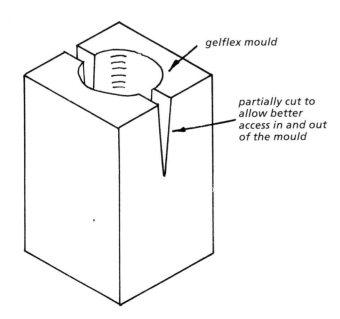

gelflex mould

partially cut to allow better access in and out of the mould

Diagram 1

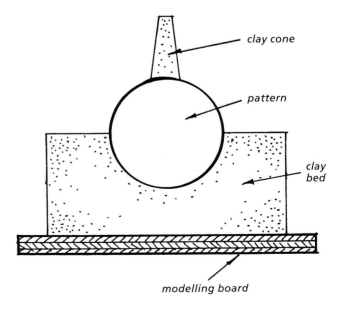

clay cone

pattern

clay bed

modelling board

Diagram 2a

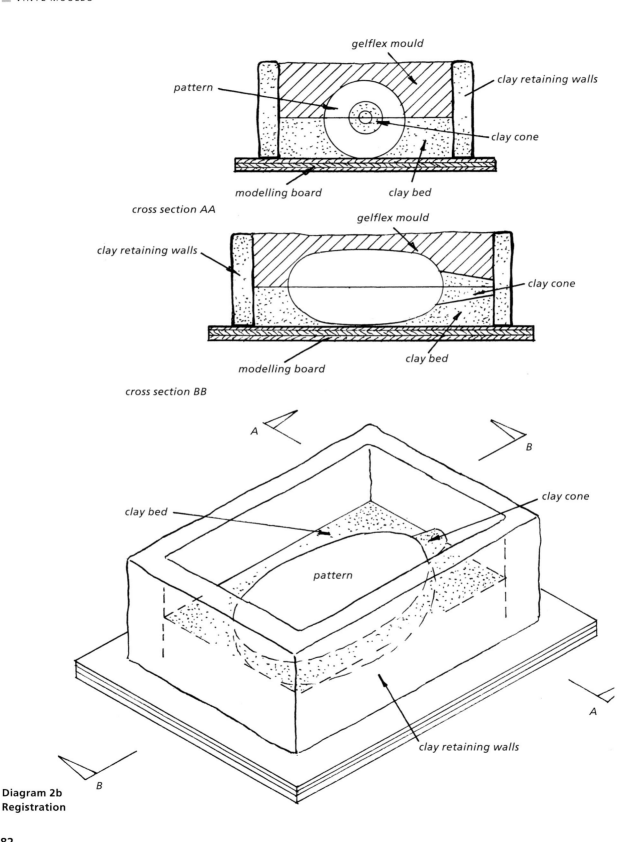

gelflex mould

pattern

clay retaining walls

clay cone

modelling board

clay bed

cross section AA

clay retaining walls

gelflex mould

clay cone

modelling board

clay bed

cross section BB

A

B

clay bed

clay cone

pattern

A

clay retaining walls

B

Diagram 2b
Registration

Gelflex will form around the cone, leaving behind a pouring hole when the cone is removed. Create the pouring hole on the part of the mould with least detail. Patterns with a wide enough foot or base can be placed on their side while embedding, with a board attached flush to the base. The mould will be made up to the base board, providing the mould mouth (*see* diagram 3).

7. Pour the Gelflex in as for the one-piece mould and allow to cool. The first piece of the mould is now complete and the second part can be created.

8. Remove the retaining wall carefully and invert the clay bed. It is important at the stage to keep the pattern within the first mould piece inverted, and it will be easiest to do this by inverting the mould and clay bed together.

9. Once inverted, pick away the clay bed carefully to expose the second half of the pattern within the first mould half. It may be necessary at this stage to clean any clay stains from the pattern before constructing the next retaining wall.

10. A light dusting of talcum powder applied to the inside face of the first mould piece will ease the separation of the two halves of the mould when complete.

11. Create a retaining wall around the first mould half to the dimension of the exposed second half of the pattern and pour the Gelflex as before.

12. Allow the Gelflex to cool, then open the mould.

OPENING AND CLEANING

1. Remove the clay walls.
2. Carefully peel the two halves of the mould apart.
3. Wash the mould in warm water.

N.B. Two-part Gelflex block moulds can be secured together for casting with rubber bands or masking tape. Do not over tighten as this will distort the mould and compromise the casting (although this can be quite interesting!).

Case Moulds

Gelflex is not suitable for large block moulds, as the unwieldy nature of a large block of Gelflex would make accurate repositioning difficult. It is therefore necessary with larger patterns to make a case mould, consisting of a skin of Gelflex approximately 10–20mm thick supported in a rigid plaster or fibreglass case. The rigid case supports the Gelflex in the original pattern's shape while casting, but can be removed and the Gelflex peeled away when the casts are ready to be removed from the mould. Using Gelflex as a skin means that very large moulds can be created economically and with maximum flexibility.

Moulds can be made in one piece if the foot of the mould is large enough to provide access into the mould; otherwise a multi-piece mould will need to be made (*see* Chapter 2).

One-Piece Moulds

Making the case is the first stage of the process. It needs to be created around the pattern, and is then filled with liquid Gelflex, which sets to form the mould. Patterns should be prepared as for other Gelflex moulds, depending on the material to be moulded, and secured if necessary to the work surface (*see* previous Gelflex moulds).

The space between the case and the pattern where the Gelflex will be poured is created using slabs of clay, temporarily laid over the pattern to form a blanket of clay, on top of which the case can then be made.

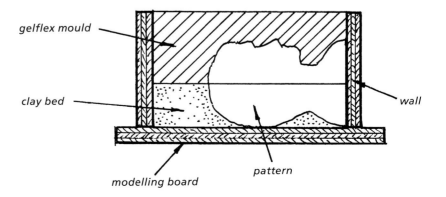

gelflex mould

clay bed

wall

pattern

modelling board

Diagram 3

1. Before laying up the clay slabs, cover the pattern with cling film or plastic to retain the moisture and prevent the slabs sticking to the pattern.
2. With a clay harp cut enough slabs of clay of approximately 15–20mm thick to cover the pattern. The width of the slabs should be between 30–60mm, depending on the size of the pattern. Slabs wider than this will be difficult to handle.
3. Gently lay the slabs over the pattern, then flat onto the modelling board or workbench, creating a flange of approximately 100mm all the way around the pattern. This will allow for registration dovetails to be cut at a later stage
4. Do not push the slabs too tightly onto the pattern; it is also important not to push them into any undercutting. If the slabs are pushed tightly into undercuts it will not be possible to remove the case from the Gelflex while casting from the mould, thus compromising the full flexibility of the Gelflex (see diagram 4).
5. Smooth the slabs together, ensuring there are no gaps, to form one harmonious, smooth blanket.
6. The next consideration is to decide the point at which the Gelflex will be poured into the mould case. This will be provided by attaching a truncated (wider at the base than the top) cone of clay at the highest point on the surface of the clay blanket (see diagram 5). When removed from the set plaster case, this will become the opening into the case through which the Gelflex will be poured. It is sometimes necessary to add smaller cones, known as risers, on larger flat moulds to allow air to escape while the Gelflex is rising in the case. A bridge between risers and the pouring point may also be necessary to allow the Gelflex to flow freely throughout the case (see diagram 6).
7. Registration dovetails should now be cut into the clay flange. These should be equidistant all the way around and will provide points of registration and anchors holding the Gelflex into the case while casting (see diagram 7). An additional location groove can also be created by laying a half-round coil of clay around the outside edge of the blanket (see diagrams 5 & 7).
8. The mould case can now be constructed. It is useful to lay a retaining wall 25mm high and 25mm out from the ends of the dovetails to contain the plaster as it is applied. Lay the case up in layers with a brush and reinforce it with scrim or fibreglass to the thickness required. Allow it to set and harden, depending on the type of plaster and the size of the case (see 'Laminating Moulds' steps 2–6, p.41).

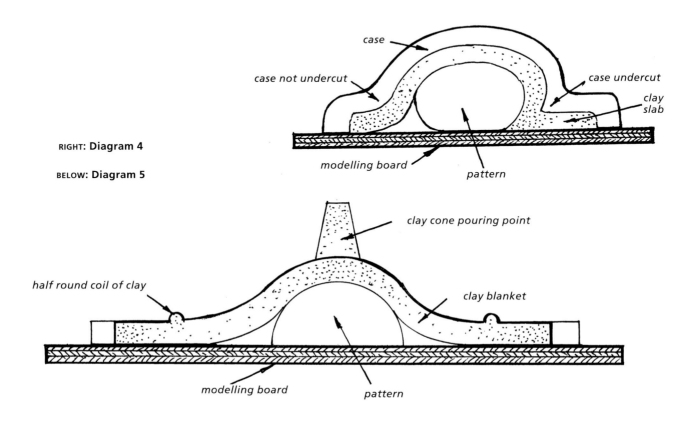

RIGHT: **Diagram 4**

BELOW: **Diagram 5**

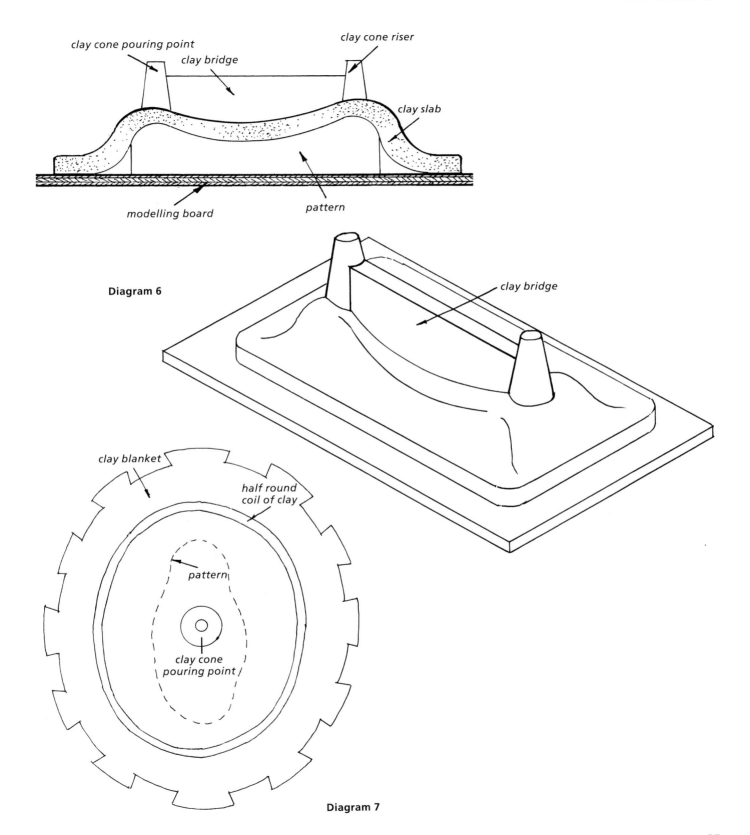

clay cone pouring point

clay bridge

clay cone riser

clay slab

modelling board

pattern

Diagram 6

clay bridge

clay blanket

half round coil of clay

pattern

clay cone pouring point

Diagram 7

9. The case is now ready to be removed, but first its position on the modelling board must be marked so that it can be repositioned accurately. Remove the retaining clay wall and mark all the way around the case; also mark a pointer at the same point on the modelling board and case as a quick reference point for repositioning.

10. Initially lift the case a little at one point and look under it to make sure that the pattern stays in position on the board as the case is lifted. This is important because if the pattern is moved while the case is being lifted, when the case is replaced it may touch the pattern, creating a hole in the Gelflex when poured. If the pattern remains in the case while lifting, tap the case until it is free and can be lifted, leaving the pattern in exactly the same position as when the case was made.

11. Remove the clay blanket. All the clay should be removed from the case, including the clay used to create the pouring hole into the mould. It is a good idea to put the clay to one side in a pile as this will give you an indication of how much Gelflex to melt to fill the case when repositioned.

12. Trim the case of any loose or weak plaster that may break off inside when the Gelflex is poured. Work around the inside of the case with a knife or chisel, removing any loose plaster. Smooth any rough or vulnerable sharp areas with a surform.

13. Thoroughly rinse the case of any plaster or clay bits.

14. Carefully remove the cling film from the pattern and remove any loose debris that may have accumulated on the modelling board while making the case. It is possible at this stage to do any final remedial repair work to the pattern.

15. Reposition the case over the pattern exactly to the line drawn on the modelling board. Secure the case to the modelling board either with screws or weights to prevent it rising when the Gelflex is poured.

16. In order to allow the Gelflex to run quickly through the case it is a good idea to extend the height of the pouring hole temporarily. This can be done by cutting the bottom out of a large paper cup and securing the cup to the pouring hole with some clay. Plug any small holes in the case with clay to prevent leaks of Gelflex.

17. The mould is now ready to use. Ensure that you have melted enough Gelflex to fill the mould, preferably in one operation to avoid chill lines (see 'Block Moulds' for pouring details). When filling a case mould, fill to the top of the paper cup used to extend the pouring hole until the case is full. Keeping the cup filled up to the top as the Gelflex is poured will provide a head of Gelflex that will push it through the case quickly before it has a chance to set.

18. Once filled, allow the Gelflex to set fully and to cool before attempting to open the mould.

N.B. A fibreglass case can be used instead of plaster. This is particularly desirable when moulding large patterns due to its light weight. *See* Chapters 3 and 8 for procedures.

OPENING, TRIMMING AND CLEANING

1. Remove any screws or weights used to secure the case to the modelling board.

2. With a slotted screwdriver or chisel gently lever the case away from the board and off the Gelflex. Ideally the case should come away from the pattern, allowing the Gelflex to be peeled away from the pattern. In some circumstances, the Gelflex may lift off with the case, in which case it should be removed once lifted.

3. Release all the dovetails from the case and ease out the Gelflex. It is important to be able to remove the Gelflex mould from its case in order to utilize the full flexibility of the compound.

4. It may be possible to remove patterns made in a soft material by digging them out, but remember that when casting in a more rigid material it will be necessary to peel the rubber away from the cast and you will need to be able to remove the case to do this.

5. Sometimes it is possible to make small cuts into the Gelflex to aid the removal of patterns and casts; such cuts will be held together in the case while casting.

6. Once the pattern has been removed, clean the mould with warm water.

ONE-PIECE CASE MOULD

The pattern is soaked in water and wrapped in cling film.

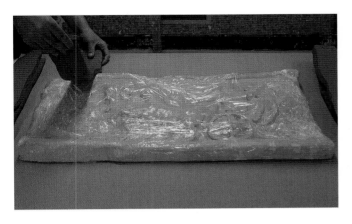

Laying on the clay slabs.

A clay blanket is made up by laying clay slabs side by side.

The slabs are smoothed together.

Dovetails are cut into the clay blanket.

A retaining wall and truncated clay cones are put in place.

A plaster case is created.

ONE-PIECE CASE MOULD *continued*

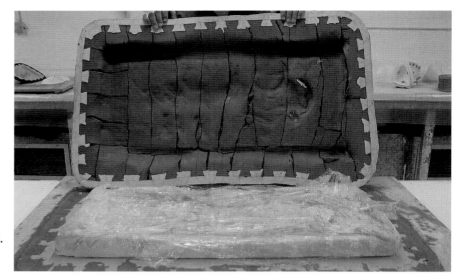

Draw a line around the case to relocate it. The case and slabs are removed from the pattern.

Removing the slabs from the case.

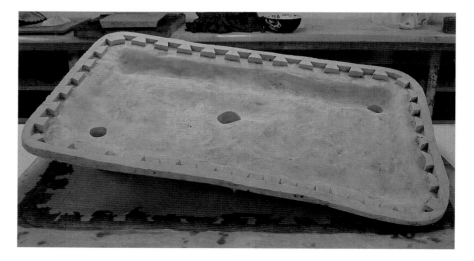

The case is empty and cleaned.

The case is repositioned to the original line.

The case is screwed down to the modelling board, the outside edge sealed with clay.

Pouring holes are extended with paper cups and the Gelflex is poured.

Excess Gelflex is cut back from the pouring points.

ONE-PIECE CASE MOULD *continued*

Cut-back excess on pouring points.

The case is removed from the Gelflex.

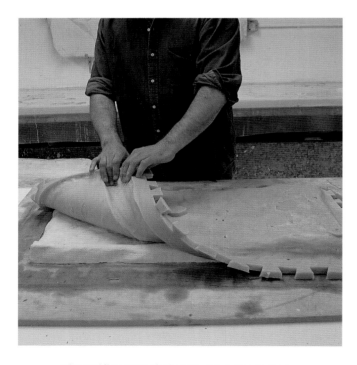

The Gelflex is peeled away from the pattern.

The completed mould. The Gelflex is back in the case ready for casting.

Multi-Piece Moulds

If the pattern to be moulded is in the round, has too small a base footprint, or has considerable undercutting, then a multi-piece mould will need to be created. A Gelflex mould of two or more pieces can be made by using either the clay bed or the wall method to divide and mask the pattern (it is not possible to use brass shim walls when making a multi-piece Gelflex mould). Both wall and bed methods are ways of dividing the pattern into sections that will become the separate pieces of the mould. The first job when making any piece mould is to determine the lines of division (*see* Chapter 2 for mouldmaking principles).

CLAY BED MOULDS

1. Wrap the pattern in cling film as described above.
2. If the pattern has a footprint, secure it to a baseboard approximately 150mm larger all round than the base. The mould will be built up onto the baseboard, eventually forming the mouth through which the casting material can be poured. The baseboard will need to be relocated as each piece of the mould is made (*see* photo sequence to follow). If the pattern is completely in the round, then a pouring hole into the mould will need to be created by attaching a truncated cone of clay at a point on the pattern with little detail. The mould will be built up around this cone, which, when removed, will become the pouring hole into the mould (*see* diagram 2a).
3. Lay the pattern down on a modelling board with the first section to be moulded uppermost and secure with lumps of clay to stop any movement.
4. Initially build up small pieces of clay onto the board around the pattern and up to approximately 5–10mm below the first section of division lines, and to a width of approximately 150mm out from the pattern. This will form a rough bed of clay around the pattern.
5. Smooth off the last 5–10mm of the clay bed up to the division line to create a smooth surface. The surface of the clay bed may undulate according to the division line. It is not necessary at this stage to take the bed tightly to the division line.
6. Cover the clay bed in cling film to prevent the slabs sticking.
7. With a clay harp cut enough slabs of clay approximately 15–20mm thick to cover the pattern. The width of the slabs should be between 30–60mm, depending on the size of the pattern. Slabs wider than this will be difficult to handle.
8. Gently lay the slabs over the pattern and flat onto the clay bed, creating a flange of approximately 100mm all the way around the pattern. This will allow for registration dovetails to be cut at a later stage. The slabs should extend up the baseboard if fitted.
9. Do not push the slabs too tightly onto the pattern; it is also important not to push them into any undercutting. If the slabs are pushed tightly into undercuts it will not be possible to remove the case from the Gelflex while casting from the mould, thus compromising the full flexibility of the Gelflex (*see* diagram 4).
10. Smooth the slabs together, ensuring there are no gaps, to form one harmonious, smooth blanket.
11. The next consideration is to decide the point at which the Gelflex will be poured into the mould case. This will be provided by attaching a truncated cone of clay at the highest point on the surface of the clay blanket (*see* diagram 5). When removed from the set plaster case, this will become the opening into the case through which the Gelflex will be poured. It is sometimes necessary to add smaller cones, known as risers, on larger flat moulds to allow air to escape while the Gelflex is rising in the case. A bridge between risers and the pouring point may also be necessary to allow the Gelflex to flow freely throughout the case (*see* diagram 6).
12. Registration dovetails should now be cut into the clay flange. These should be equidistant all the way around and will provide points of registration and anchors holding the Gelflex into the case while casting (*see* diagram 7). Dovetails should also be cut into any slabs extending onto the base board if fitted. An additional location groove can also be created by laying a half-round coil of clay around the outside edge of the blanket (*see* diagrams 5 & 7).
13. Before laying up the plaster case, registration points should be created in the clay bed approximately 20mm out from the ends of the registration dovetails. Using a round-ended tool handle, they should be shallow but large enough to provide positive registration between the case pieces.
14. The mould case can now be constructed. It is useful to lay a retaining wall 25mm high and 25mm out from the ends of the dovetails to contain the plaster as it is applied. Lay the case up in layers with a brush and reinforce it with scrim or fibreglass to the thickness required. Allow it to set and harden, depending on the type of plaster and the size of the case (*see* Chapter 4, 'Laminate Moulds'). Alternatively, a fibreglass case can be made (*see* 'Fibreglass Moulds').
15. Keeping the case containing the clay blanket and the pattern inside together, carefully invert the whole structure including the clay bed and secure it to the modelling

board so that it cannot move around. It is important during this procedure to keep everything together as it was constructed, otherwise you may have problems with the registration of mould pieces later. Carefully pick away the clay bed; the cling film will aid this procedure by exposing the next section of the pattern to be moulded.

16. The pattern will now be laying in the clay blanket in the plaster case. Due to the way that the clay slabs have been laid over the pattern and slightly flared out, when the structure is inverted there will be a gap between the blanket and the pattern. If the mould is to be made in two pieces this gap can now be filled with more clay, effectively creating a new clay bed exposing the next section of the pattern. When filling this gap, take the clay tight up against the pattern using a clay tool to keep the bed at right angles to the pattern (*see* diagram 8). It is important to keep the bed at right angles to avoid feathered (weakened) edges to the mould pieces that would make them vulnerable to damage. The clay bed at this stage should not be higher than the level of the clay slabs sitting in the plaster case. If the mould is to be made in more than two pieces it is not necessary to fill to the pattern tightly at this point; this will be done later when opening the case piece by piece and preparing for pouring the Gelflex.

17. With the bed in place, lay cling film over the pattern, clay bed and exposed plaster case to prevent the clay slabs from sticking. Lay the clay slabs up as before, smoothing into one blanket and cutting registration dovetails. The dovetails should extend the same distance as before, but do not have to match up exactly with the ones underneath.

18. The second piece of the plaster case can now be made. The cling film should cover the exposed surface of the first plaster case, which will prevent the new plaster sticking to it, but if there is any exposed plaster apply a little petroleum jelly as a release agent.

19. Make a clay wall around the outside of the first case to retain the first layers of plaster laid up on the second and prevent plaster running down and sticking the two together. Lay the plaster case as before and allow to set and harden fully. Remember to apply release agent between case pieces as they are made.

20. If the mould is to be made in more than two sections repeat the previous procedures for each section of the mould, remembering always to keep the case pieces together and the pattern in position as you move on to the next piece of the case to be created. When all the case pieces have been created, preparation for pouring the Gelflex is the next stage.

CASE PREPARATION AND POURING THE GELFLEX

1. With all the case pieces now complete, consideration should be given to the method of securing the mould pieces together while it is being cast in. Depending on size, there are several ways to hold the mould pieces together (rubber bands, cut up pieces of bicycle inner tube, clamps or nuts and bolts). If using nuts and bolts, holes should be drilled (any sort of drill bit can be used and should be as close to the nut and bolt size as possible) before the case parts are separated to ensure tight registration of pieces when reassembling.

2. Secure the mould with lumps of clay or bricks to the modelling board with one case piece uppermost.

3. Insert a strong, thin-bladed tool between two case pieces and lever them gently apart. It is important to leave the pattern sitting in the other case piece or pieces when removing the case piece to be prepared. Once removed, the case piece should be emptied of clay slabs, trimmed and cleaned (*see* steps 11–13 in the one-piece method above).

4. If not previously filled, the gap between the clay blanket and the pattern should now be filled using a modelling tool, ensuring that there are no gaps through which the Gelflex can leak and that the bed is at right angles to the pattern (*see* diagram 8).

5. The prepared case piece can now be put back in position and secured to its adjoining case pieces. Extend the Gelflex pouring hole with a paper cup, pour the Gelflex, and allow it to set thoroughly and cool (*see* step 17 in the one-piece method above).

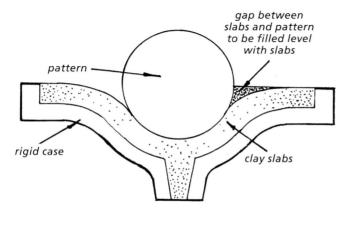

Diagram 8

gap between slabs and pattern to be filled level with slabs

pattern

rigid case

clay slabs

6. With the first section complete, rotate the mould round to the next section of the case to be filled and stabilize it on modelling board. Prepare, reassemble and pour Gelflex as previously described. Take note that when the next section of the case has been removed you will see that the previously poured and set Gelflex forms the bed that the pattern sits in (or part of the bed in a mould of more than two pieces). A dusting of talcum powder on the exposed Gelflex will aid parting of the adjoining Gelflex piece when opening the mould.

7. Methodically moving around the mould, create each section against the next, following the same procedures, until all the case sections are filled and the mould complete.

OPENING, TRIMMING AND CLEANING

Before opening the mould it is important to consider how the pieces of the mould will be held together while casting. With the smaller moulds it may be adequate to use large rubber bands or cut-up lengths of bicycle inner tube, tied and made into bands of appropriate lengths. For the larger moulds the mould pieces can be bolted together – this is where a wide flange is vital. A wide flange will provide a large surface area of registration between mould pieces, but also plenty of space to drill holes to take a nut and bolt. If using nuts and bolts to secure mould pieces, all the holes to take them should be drilled before opening the mould to ensure accurate reassembly (this will only be necessary if a secure system has not been put in place at case preparation stage). Any hand or electric drill with a standard HSS (high-speed steel) twist drill bit can be used to do this. Bolts with wing nuts should be used for ease of reassembly, and a large washer on each side of the bolt (all should be galvanized so as not to rust). The mould can now be opened, trimmed and cleaned.

1. All flanges should initially be cut back a few millimetres to expose the seam between them and provide a strong, tidy edge. Using a surform, take the excess plaster back until a distinct line can be seen between the case pieces.

2. Once all the seams have been exposed, remove the case pieces. Start by inserting a thin, strong-bladed knife in-between all the case pieces. Initially, this is just to part the adjoining flange seams of each piece, not to remove the pieces from the Gelflex.

3. Once the flanges have been parted, open up the seams some more by methodically working equally and gradually all the way along them. As the seams start to open it may be necessary to use a chisel, particularly on larger moulds. Larger moulds may also benefit from a series of wooden or plastic wedges being gradually driven into the seam in sequence until the mould pieces separate. Once one case piece has been removed the others will generally come away easily.

4. Once all the case pieces have been taken off, remove the Gelflex sections of the mould. Gently peel them away from each other and the pattern.

5. Trim the case pieces and sand them for ease of handling.

6. Wash the mould in warm water, and dry it by hand.

7. Reassemble the mould securely to prevent it warping before it is used.

MULTI-PIECE CASE MOULD

The original pattern.

MULTI-PIECE CASE MOULD *continued*

The pattern is wrapped in cling film, laid down, the base board secured and the clay bed created to mask one half.

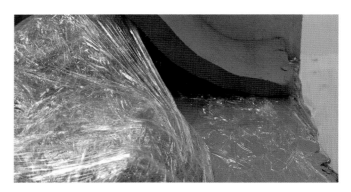

The cling film is laid on the clay bed to prevent the clay slabs sticking.

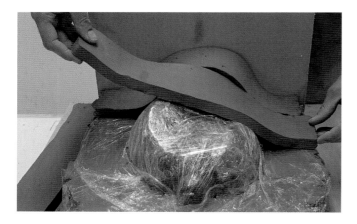

The clay slabs are laid on and smoothed together.

Dovetails are cut, truncated cones placed and a retaining wall built.

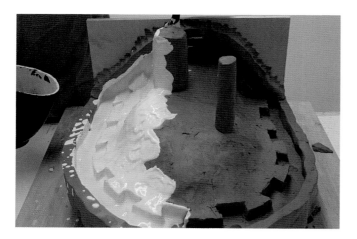

A plaster case is built up in laminated layers.

The mould is inverted and the clay bed removed, exposing the other side of the pattern.

The slabs and the case are created for the second half of the mould.

The case and the slabs are removed from one half of the pattern.

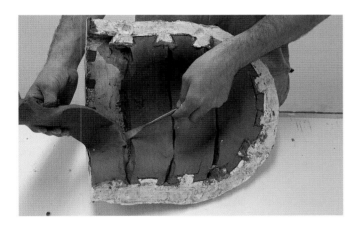

Removing the slabs from the case.

The gap between the pattern and the slabs is filled.

The case, now free of slabs, and the base board are clamped back, the pouring point is extended and Gelflex is poured.

The set Gelflex mould is inverted, and the second piece is removed to be prepared for next Gelflex pour.

MULTI-PIECE CASE MOULD *continued*

WALL DIVISION LINES

When making multi-piece Gelflex moulds it is also possible to create pattern divisions for the mould pieces using walls instead of a clay bed. This can be advantageous, for instance when moulding a large standing figure that would be difficult to lay down horizontally and would require very large amounts of clay to create a bed.

1. The same considerations need to given to undercutting and pattern divisions as for clay bed construction and the mould pieces are created in the same way, one by one, each connecting to build up the mould.
2. Walls are used to mask the sections of the pattern as each piece of the mould is made. The wall material can be clay or any other material that can be temporarily but securely attached to the pattern as the pieces are constructed.
3. They should be attached as perpendicular to the pattern surface as possible; clay slabs are then laid up in the same way as for a clay bed (*see* diagrams 9a and b).
4. Consideration should be given to the Gelflex pouring point, which needs to be at the highest point on the slabs. Because you may be working on a pattern that is upright, the mould sections will be constructed on a vertical plane, which means that the truncated cone placed on the slabs should be at the highest point to avoid trapping air when the Gelflex is poured (*see* diagram 9c).
5. Cases should be constructed using the laminating technique, as they will be larger and likely to be built vertically. The Gelflex should be poured as for the clay bed method.
6. It is advisable to pour each section of Gelflex once each piece of the case has been made when using the wall method, rather than construct all the case pieces first and then pour each section of Gelflex as with the clay bed method.
7. When moving onto the next section of the mould, remove the clay walls and rebuild them, defining the next section of the pattern to be moulded (*see* diagram 9d). The steps described are then repeated until the mould is complete.

TOP: **The mould complete, holes are drilled for nuts and bolts before demoulding.**

MIDDLE: **The case is removed and the Gelflex is peeled off the pattern.**

BOTTOM: **The completed mould.**

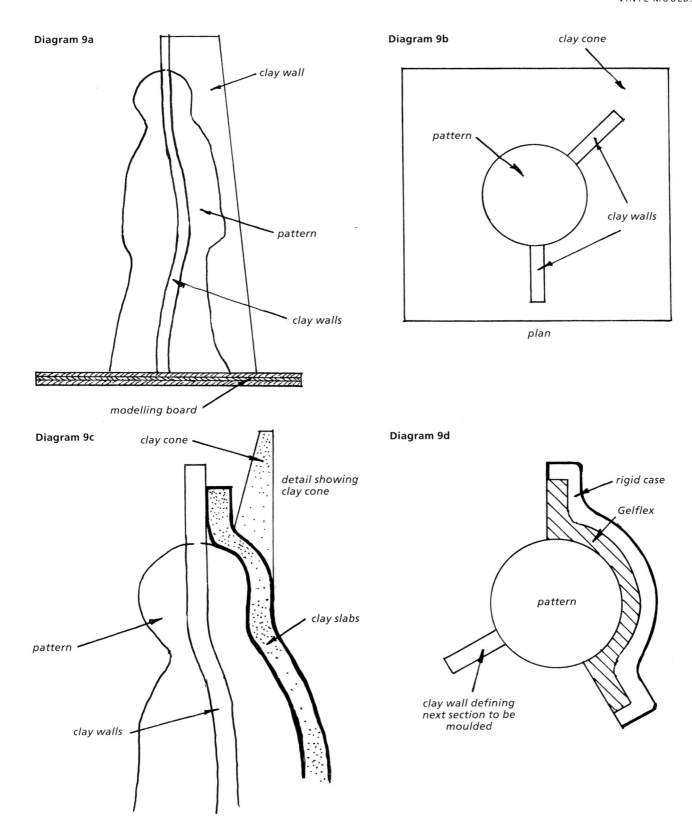

Diagram 9a

clay wall

pattern

clay walls

modelling board

Diagram 9b

clay cone

pattern

clay walls

plan

Diagram 9c

clay cone

detail showing clay cone

pattern

clay slabs

clay walls

Diagram 9d

rigid case

Gelflex

pattern

clay wall defining next section to be moulded

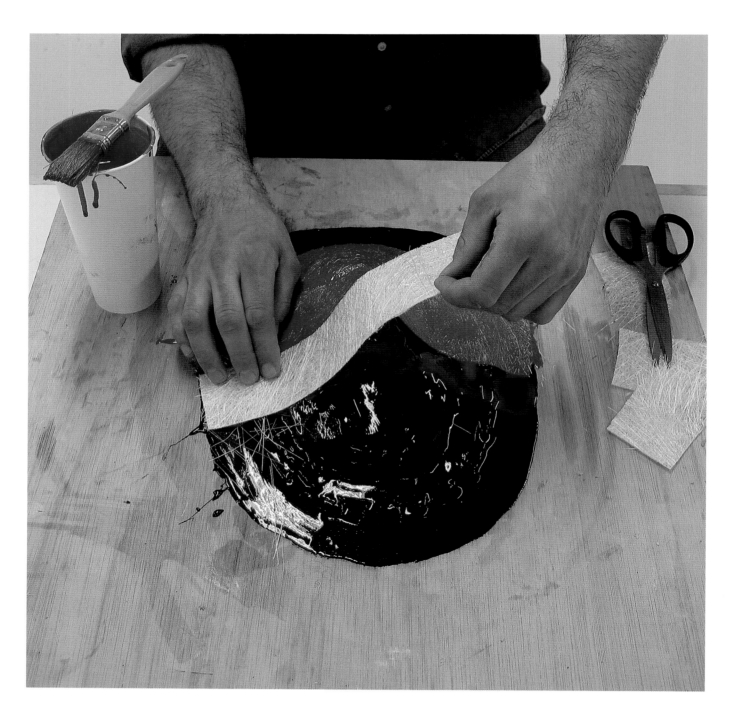

Laying fibreglass mat onto the gel-coat of a mould.

CHAPTER 8

FIBREGLASS MOULDS

Fibreglass and polyester resin as mouldmaking materials can produce durable, lightweight yet rigid moulds with high degrees of detail capture. Polyester resins are liquids of various viscosities that set to a rigid plastic with the addition of a catalyst. When applied to a pattern they will run into the detail and set hard; fibreglass mat is then applied to add strength, forming the mould. Note: It is also possible to make moulds with polyester resin paste (car body filler).

Being a rigid moulding material, if patterns have undercutting moulds will need to be made in pieces in order for it to be possible to remove them from patterns and subsequent castings. As a very light material, fibreglass moulds are suitable for very large patterns, where the weight of other moulding materials might be a problem.

One-piece or multi-piece moulds can be made using fibreglass. Specifications of the material, tools and the principal techniques of using polyester resins and fibreglass should be read before embarking on mouldmaking. Fibreglass moulds can be taken from a number of different pattern materials, the type of which will determine whether or not a release agent will be required (*see* Chapter 3). If making multi-piece moulds, pattern undercutting, mould division lines and the method of masking the divisions (clay bed or walls) will need to be determined. These considerations are common to other rigid moulding processes and should be read in Chapter 2 before mouldmaking.

For materials, tools and specific techniques, *see* Chapter 11.

One-Piece Moulds

It is only possible to make one-piece moulds from a pattern with no undercutting, otherwise the pattern will be locked into the mould. This means that the footprint (base) of the pattern by definition will need to be wider than the top (for example, conicular or pyramidal). The footprint of the pattern in its negative moulded form will become the mouth of the mould.

Patterns should be firmly secured to a modelling board and any gaps between its base and the board should be filled with plasticine to prevent resin from seeping underneath. Plasticine should be used rather than clay to fill any gaps as a more efficient release agent can be applied to it. The relevant release agents for the type of pattern material and any plasticine gap fillings should now be applied. Release agent should also be applied to the modelling board, 100mm out from around the base of the pattern, as the mould will be built up against the board surface at this point.

Making a fibreglass mould is principally a two-stage process involving two types of polyester resin, gel coat and general purpose (GP), and a reinforcing laminate fibreglass.

Gel Coat (First Layer)

1. Apply a layer of gel coat resin, catalysed at 2 per cent, by brush to the entire surface of the pattern and onto the surface of the modelling board out to 100mm around the base of the pattern. Take care to ensure that the resin is brushed into all the pattern detail, as this will be the surface of the mould, to a depth of approximately 2–3mm.

2. Allow the gel coat to set for 30–40 minutes until firm to the touch but still tacky. If there is a lot of fine raised detail on the pattern surface it may be necessary to apply a second gel coat at this point. Application of a second gel coat should be as for the first and it should be allowed to set as before.

Fibreglass Laminate (Second Layer)

1. Cut enough pieces of 450g fibreglass mat to cover the whole surface of the previously applied gel coat. The size and shape of the pieces will be dependent upon the size and shape of the pattern, and what will most efficiently laminate the gel coat.

2. Dispense and catalyse at 2 per cent enough GP resin to use within its pot life, and apply a layer by brush to the first area to be laminated. It will be necessary on larger moulds to dispense and catalyse several batches of GP resin, otherwise it will have started to set before it can be applied. If a batch starts to gel, throw it away and use another batch, as it will not brush out evenly and soak into the mat.

3. Lay a piece of fibreglass mat onto the wet resin and then saturate thoroughly with more GP resin. Do not over-brush the mat as it will bring up the fibres of the mat.

4. Repeat steps 2–3 on an adjoining area of gel coat, overlapping the mat by a few millimetres. While you are doing this, the GP resin in the first piece of mat will be soaking in and softening it.

5. Go back to this first piece of mat now, and with the tips of the brush bristles stipple the whole surface, pushing the air out and ensuring a good lamination between gel coat and fibreglass.

6. Repeat this process until the whole surface of the gel coat has been laminated with mat.

7. Allow the mould to set thoroughly before opening.

N.B. If a mould with slight flexibility is required a lighter weight mat can be substituted. Larger moulds or moulds that require more rigidity may require two layers of mat or a heavier weight mat. Very large moulds may require reinforcement with splints of timber or steel attached to the outside of the mould with more mat and resin.

OPENING, TRIMMING AND CLEANING

1. Insert a thin, strong-bladed knife between the mould and modelling board and methodically release all the way around.

2. Carefully lever the mould away from the modelling board. Gloves should be worn to avoid any sharp fibreglass strands on the body of the mould.

3. A combination of tapping the mould and levering out the pattern should release it from the mould. If the pattern will not release, soaking it in hot soapy water or under a running warm tap should help. Bear in mind that the pattern may be damaged during these procedures.

4. Once released, trim the pattern. The flange of fibreglass that was in contact with the modelling board, which is now the mould mouth, should be cut back a few millimetres to a strong line, and the body of the mould sanded with wet and dry paper to make it safe to handle.

5. Wash the mould thoroughly with hot water and detergent. It is now ready for casting.

Multi-Piece Moulds

Any pattern with undercutting will require a mould made in pieces in order to be able to remove the pattern and subsequent castings. Fibreglass moulds, even when used in their lightweight form, will usually not be flexible enough to bend round undercutting.

Multi-piece moulds can be made around patterns that are completely in the round, that is, they have no footprint (base); however, consideration will need to be given to allow an opening into the mould through which the casting material can be introduced. A pattern with a footprint will have the mould constructed around it, leaving an opening into the mould. Patterns in the round will need an opening created in one of the pieces to allow access. This is done by attaching a temporary 'stalk' to the body of the pattern, around which the mould is made. When the mould is complete and the stalk removed, an opening into the mould is left.

When making a multi-piece mould, the division of the pieces should occur along the high points of any undercut area. The division lines should be determined and either marked on the pattern surface or committed to memory.

The next consideration will be the method of masking each area of the pattern as it is made. Each piece of the mould will need to be made separately and the other areas of the pattern that the mould pieces are to be made from will need to be masked while each piece is made. This can either be done with a temporary clay bed or with walls that are built up to the division lines. The mould piece is then made, the bed or wall removed and reconstructed along the next set of division lines defining the next piece to be made.

Deciding whether to use a bed or walls will largely depend on the size of the pattern. Due to the amount of clay involved the clay bed method is more suitable for smaller patterns and walls more suitable for larger ones.

Clay Bed Method

1. Secure the pattern temporarily to a modelling board to stop it moving while making the mould. It may be advantageous to lay taller or larger patterns down horizontally. Place the pattern so that the first piece to be moulded is uppermost.

2. If the pattern has a base, a base board should be temporarily secured flush to it. This will mask off the base, the mould being built upon it forming the mouth of the mould. If moulding a pattern completely in the round, a cone of plasticine should be attached to the pattern. When the mould is complete and the stalk removed it will leave a hole through which the casting material can be poured. Stalks should be positioned on a section of the pattern where the subsequent casting can be easily reworked, as there will be a loss of detail at this point. The stalk diameter will depend on the size of the pattern and what casting material will be poured into the mould.

3. Start to build clay up to the first set of division lines on the pattern. Initially use large chunks of clay, getting smaller as you approach the lines on the pattern.

4. When up to the division lines use a clay modelling tool to bring the clay bed tight to the pattern. The clay bed should be made as perpendicular to the pattern as possible to avoid feathering (weakening) of the mould edges.

5. The clay bed should extend at least 150mm out from the pattern to create a flange. The flanges around the edges of the mould pieces will create a positive connection between the pieces.

6. Apply release agent to the pattern, clay bed and base board, if there is one.

7. The first piece of the mould can now be created. Follow the steps as for the one-piece mould (gel coat and fibreglass, steps 1–6) above. The mould should be built out at least 50mm onto the clay bed to provide a wide flange around the mould piece edges. This will allow for positive reassembly of the mould, providing an area to clamp or bolt the pieces together. Allow the first piece of the mould to set thoroughly.

8. The next piece of the mould can now be made. Remove the pattern from the clay bed, ensuring that the first mould piece remains in situ on the pattern. The next section of the pattern to be moulded should be selected next to the first mould piece, and the pattern secured in its new position. If a base board is being used this will need to be repositioned as well.

9. A new clay bed can now be made, defining the next piece of the pattern to be moulded. A section of the next mould piece will connect with a section of the first mould piece. At this section, the clay bed should be built up tight to the first mould piece.

10. Follow steps 5–6 again, creating the next piece of the mould.

11. Continue in this way, methodically moving from one section of the pattern to the next, to create as many mould pieces as necessary to complete the mould.

Wall Division Lines

1. This method uses walls to mask off the sections of the pattern to be moulded. Usually clay is used as the wall building material as it is flexible and will adhere to most materials temporarily. However, plasticine is preferable when using resins as a more efficient release agent can be applied. Theoretically, any material could be used (wood, card and so on), as long as it can be fixed tightly to the pattern with no gaps for resin to seep through, and can be removed cleanly without damage to the pattern.

2. Rigid materials will need to be cut approximately to the contours of the division lines on the pattern, and will usually need to be sealed with plasticine at the final point of contact. Bear in mind that this method will usually be applied to patterns that may be too large to move; walls may therefore have to be built onto a vertical surface so they will need to be fixed firmly in place during the mouldmaking process.

3. The width and thickness of the walls will largely depend on the size of the pattern, but should be of a width to incorporate a large flange (at least 50mm) around each mould piece, so as to provide a wide surface area and fixing point between mould pieces.

4. As when creating a clay bed, keep the edges of contact perpendicular to the pattern on the high point of any undercutting. When creating the mould pieces this will ensure that the mould edges are not feathered (thinned), and that the pieces are slightly plug-shaped so that they will come apart easily.

5. If moulding a standing pattern that has a base, the walls will need to be taken down to the modelling board for each section. Build the pieces of the mould against the modelling board at these points, creating the mouth of the mould through which it will be filled. If moulding a pattern completely in the round, a stalk of plasticine should be attached to the pattern. When the mould is complete and the

stalk removed it will leave a hole in this piece, through which the casting material can be poured. Stalks should be positioned on a section of the pattern where the subsequent casting can be easily reworked, as there will be a loss of detail at this point. The stalk diameter will depend on the size of the pattern and what casting material will be poured into the mould.

6. Apply appropriate release agents to the walls and pattern.

7. With a first pattern section walled, follow the steps as for the one-piece mould (gel coat and fibreglass, steps 1–6) above. Build the mould out at least 50mm onto the walls to provide a wide flange around the mould piece edges. Allow the first mould piece to set thoroughly.

8. Remove the first set of walls, ensuring that the first mould piece stays in situ on the pattern.

9. A new set of walls can now be attached, defining the next section of the pattern to be moulded. The next mould piece to be made will connect with the first mould piece. A section of the walls defining the new section will be a section of the flange from the first mould piece. The walls should be built up tight to the first mould piece at this point.

10. If a mould is to be made in just two pieces, the flange of the first piece provides the wall up against which the second piece will be created.

11. Apply release agents to the new walls and exposed flange.

12. Follow the method as for step 7.

13. If more mould pieces are to be made, repeat steps 8–9 until all mould pieces are created, each piece connecting to another to complete the mould.

14. If the mould is in the round do not forget to create a pouring hole in one of the mould pieces (*see* step 5).

N.B If a mould with slight flexibility is required a lighter weight mat can be substituted. Larger moulds or moulds that require more rigidity may require two layers of mat or a heavier weight mat. Very large moulds may require reinforcement with splints of timber or steel attached to the outside of the mould with more mat and resin (*see* Chapter 7 for more information on 'walls').

OPENING, TRIMMING AND CLEANING

Before opening the mould it is important to consider how the pieces of the mould will be held together while casting.

With the smaller moulds it may be adequate to use large rubber bands or cut-up lengths of bicycle inner tube, tied and made into bands of appropriate lengths. For the larger moulds the mould pieces can be bolted together – this is where a wide flange is vital. A wide flange will provide a large surface area of registration between mould pieces, but also plenty of space to drill holes to take a nut and bolt. If using nuts and bolts to secure mould pieces, all the holes to take them should be drilled before opening the mould to ensure accurate reassembly. Any hand or electric drill with a standard HSS twist drill bit can be used to do this. Bolts with wing nuts should be used for ease of reassembly, and a large washer on each side of the bolt (all should be galvanized so as not to rust).

The mould can now be opened, trimmed and cleaned.

1. Cut back all flanges by a few millimetres to expose the seam between them and provide a strong, tidy edge. This can be done by hand or mechanically.

2. Start by inserting a thin, strong-bladed knife in-between all the mould pieces. Initially, this is just to part the adjoining flange seams of each piece, not to remove the pieces from the pattern.

3. Once the flanges have been parted the seams can be opened up more by methodically working around the seams. Open the seams equally and gradually all the way along, rather than in one place.

4. As the seams start to open it may be necessary to use a chisel, particularly on larger moulds. Larger moulds may benefit from a series of wooden or plastic wedges being gradually driven into the seam in sequence until the mould pieces separate. Once one mould piece has been removed the others will generally come away easily.

5. With all the mould pieces removed from the pattern, sand the flange edges and the outside of the mould with wet and dry paper for ease of mould handling. Do not work the inside of the mould as it will compromise the moulded surface.

6. Wash the mould in warm water and detergent, and dry it by hand.

7. Reassemble the mould securely to prevent it warping before it is used.

**'Andy'. By Steve Yeates. Resin and glass aggregate.
H 540mm × W 270mm × D 200mm.**

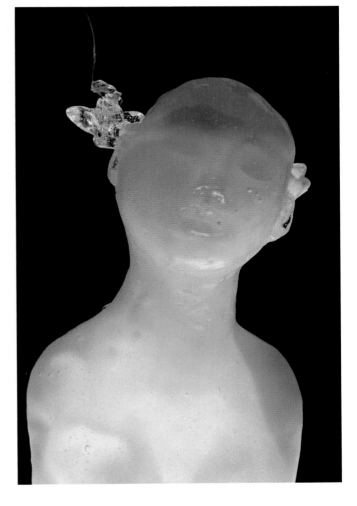

Ice sculpture. By Naoko Kitamura.

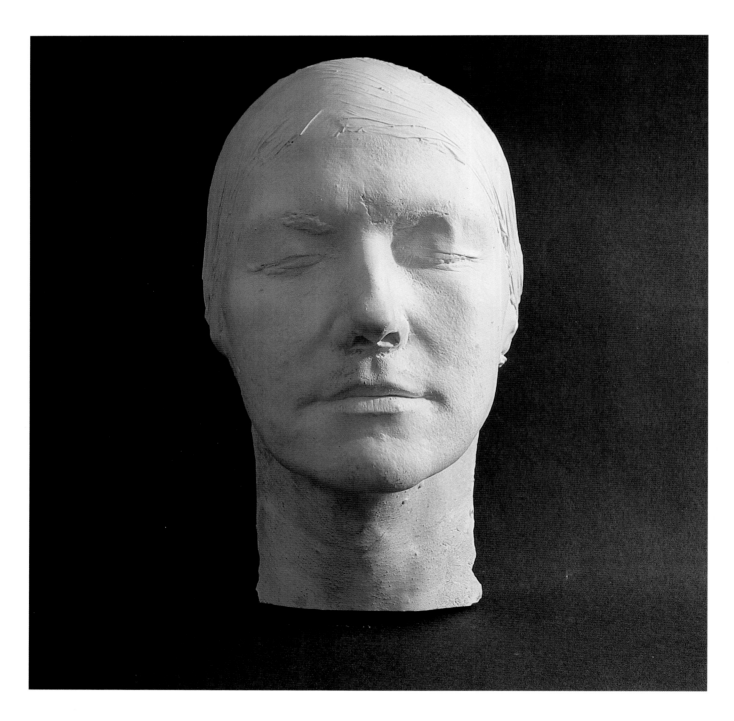

Life cast from face.

CHAPTER 9

LIFE MOULDS

Life moulding is the technique used to reproduce castings of a living person, that is, a mould is created directly from a life model. Generally, a plaster casting is taken from a life mould, as the mould will usually only survive one casting. If multiple castings in other materials are required the plaster cast can be used to remould in a more durable moulding material.

Because the mouldmaker will be working with a life model, safety is the first concern. The mouldmaker should be fully familiar with the techniques involved, initially working on a safe part of the body before attempting more complicated areas. For instance, it is strongly advisable to learn the techniques by taking a mould from the hand before attempting to do a face! There are two safe materials that can be employed, plaster bandage and alginate. Both are relatively fast setting and lightweight compared to other mouldmaking materials, which is essential when working with life models who are having to hold a pose – but the main consideration is that they are safe to use.

The Materials and Principal Considerations

Plaster Bandage

Plaster bandage is a fine cotton scrim that has been pre-impregnated with plaster; it is used in the medical industry to repair broken limbs. It is usually available in rolls of approximately 150mm × 2.5m or in 6kg bags in pieces of varying sizes. Quality may vary between batches, particularly in the bags; pieces with little plaster on should be discarded. The way that the bandage is impregnated during manufacture creates one side with more plaster, and this side should be laid against the skin of the model.

The method of application is to dip the plaster bandage briefly in water, squeeze the excess water gently out, apply the bandage to the skin of the model and allow it to set, thus forming the mould. The initial setting time is 2–5 minutes.

- **Storage:** airtight container, cool dry place, six months to one year.

Alginate

Alginate is a powder that is mixed with water to a paste which then sets to flexible solid; it is used in the dental industry to produce moulds and castings from teeth.

The paste is applied to the model's skin; once firm, it is backed up with a case of plaster bandage that will support it during casting. The initial setting time of alginate is 2–5 minutes.

- **Storage:** airtight container, cool dry place, one to two years.

Release Agents

To release either of these moulding materials from the skin either petroleum jelly or baby oil should be used. Petroleum jelly should be applied to skin and rubbed in so the surface remains greasy but the detail is not clogged with excess that would compromise the mould surface. Any 'hairy bits' should be thoroughly 'jellied' and any hairs standing up should be flattened down with jelly. Hairs left standing up will be trapped in the mould and will either have to be cut or pulled off with the mould! Be thorough with release agents; your model will thank you!

Life Model

Life moulding will inevitably be slightly uncomfortable for your model so ensure that they are as comfortable as possible before starting. If the model is to hold an awkward pose, the limbs could be supported or the model seated while the mould is being made. Use a hypoallergenic barrier cream if the model has sensitive skin. Above all, your model will want to feel that you know what you are doing and that you can work rapidly, efficiently and safely.

Plaster Bandage Two-Piece Mould From the Hand

To take a life mould of a whole hand will require a mould in two pieces, as the plaster bandage will become rigid when it sets, making it at the least difficult and at the worst impossible to remove the hand without damaging the mould.

A line defining the two halves of the mould can either be drawn around the hand or committed to memory. The line should follow the outline of the hand at approximately the halfway point, with the palm as one half and the top of the hand the other. It is not necessary for the mould to be made to exactly the halfway point as there will be a small amount of flexibility in the mould while it is still wet to aid its removal; there is also flexibility in the flesh of the model.

Preparation

1. Cut and prepare enough strips of plaster bandage for the job. The length of the strips should be the width of the hand from the division line to the division line on each side plus an extra 80mm. The extra will provide enough bandage to fold back onto itself to create a flange around each of the two mould pieces, so that they will register positively together when casting. The width of the strips should be approximately 30–50mm.
2. Cut enough strips to cover the entire hand twice, overlapping each strip. Cut an extra five or six strips to be used for reinforcing ribs.
3. Methodically lay out the strips on a dry modelling board in the order they are to be applied to the model. It is important to prepare all the bandage prior to moulding, as when you

start you will have two wet hands that will need to be dried before handling more bandage if you do not have enough; also you will need to work methodically and rapidly.
4. A bowl of cold water should be placed on the bench in a position that will not splash the prepared strips of bandage.
5. Apply petroleum jelly, rubbing it into the skin so as not to clog the detail.

First Piece

1. Determine which side of the bandage has the most plaster. This will be the side to be laid onto the skin of the model (it is sometimes easier to feel which side has the most plaster).
2. With one hand, dip the first strip to be laid up into the bowl of water for a few seconds, lift it out and then squeeze out the excess water between the index and middle fingers of the other hand. As well as expelling excess water this has the effect of passing the wet plaster evenly throughout the gauze of the bandage.
3. Lay the strip onto the skin and rub gently, pushing the bandage into the detail. Rubbing the bandage pushes the plaster through the gauze onto the surface of the skin, picking up the detail and pushing out any trapped air.
4. Continue laying up strips in the same way, in sequence side by side overlapping by 2 or 3mm. It is important while laying up the strips to leave approximately 80mm of extra bandage to fold back for the mould flange.
5. Create the flange as you go, by folding back onto itself and pinching it into a flange approximately 10–15mm wide following the halfway line around the hand. Be sure to do this as you go along as it will not be possible if the bandage has started to set. The flanges of any mould are crucial in the accurate reassembly of the mould for casting. (If the mould was to be continued to another piece incorporating the arm, a flange would need to be created around the wrist to connect the two mould pieces (see diagrams 1a and b).
6. With the first layer of bandage in place, apply the second layer in the same way, reinforcing the flange as well.
7. Put the reinforcing bars in place. Dip and squeeze out the reinforcing strips and then lay them on a clean surface. Fold the bandage into a strip approximately 15mm wide and place across the moulds width. Pinch the strip to a point, forming a triangular reinforcement bar across the width of the mould. Continue to apply bars at approximately 30mm intervals along the length of the mould.
8. The first piece of the mould is now complete.

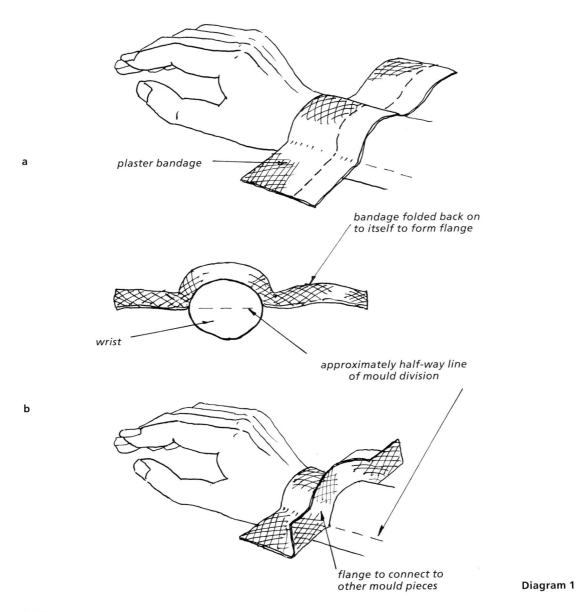

a

plaster bandage —

bandage folded back on to itself to form flange

wrist —

approximately half-way line of mould division

b

flange to connect to other mould pieces

Diagram 1

Second Piece

1. The second piece of the mould needs to be created tightly against the first to ensure a minimum seam line between the two, so it is crucial to keep the first piece in place on the model as it was created. It can be a little awkward for the model to flip the hand over to expose the unmoulded side of the hand, so allow them to find a comfortable way to do this and to reposition ready for the next piece to be made.

2. The unmoulded side of the hand will have inevitably picked up traces of plaster while the first piece of the mould was being made and these will need to be carefully cleaned off and the hand covered with petroleum jelly as before.

3. As well as jellying the skin, the flange of the first piece should be carefully jellied so that the two mould pieces will not stick together.

4. The bandage should be cut, prepared and laid-up as before. Take care to create the second flange tightly up against the first, so as to minimize the seam.

With both pieces of the mould complete, allow the last piece of bandage that was laid up to harden for 10–15 minutes before attempting to open the mould.

Opening the Mould and Casting

- Removing a life mould is a combination of the mouldmaker easing the mould off and the model easing out of it. The model will be able to gauge more easily how to get out of the mould without either personal damage or damage to the mould.
- Once free from the model, the mould can either be cast into directly or kept for casting in at a later date.
- If it is to be cast in directly, the mould pieces should treated with the appropriate release agents and be sealed together with a double strip of bandage around the seam ready for casting (*see* Chapters 3 and 10).
- If the mould is to be used at a later date the mould pieces should be secured together temporarily, with rubber bands or masking tape to avoid warping while drying out. Once fully dry (overnight in a warm dry place), the mould should be parted and release agents applied ready for casting (*see* Chapters 3 and 10).
- After the cast has been made the mould can be soaked in warm soapy water to aid removal.
- Generally one casting is all that can be produced from a life mould, the mould being wasted (destroyed) during cast removal.
  Note: It is possible to cast other materials (*see* Chapter 3).

Alginate Two-Piece Mould with Plaster Bandage Case From the Hand

To take a life mould of a whole hand will require a mould in two pieces, as the plaster bandage will become rigid when it sets, making it at the least difficult and at the worst impossible to remove the hand without damaging the mould.

A line defining the two halves of the mould can either be drawn around the hand or committed to memory. The line should follow the outline of the hand at approximately the halfway point, with the palm as one half and the top of the hand the other. It is not necessary for the mould to be made to exactly the halfway point as there will be a small amount of flexibility in the mould while it is still wet to aid its removal; there is also flexibility in the flesh of the model.

Preparation

1. All the mould materials should be prepared prior to making the mould. Enough plaster bandage should be cut and laid out as described for the plaster bandage method above. Lengths of bandage will need to be a little longer to accommodate the thickness of the alginate as well as the width of the hand.
2. For the alginate you will need a ready supply of clean cold water, waxed paper cups or a bowl for mixing the alginate in, and wooden spatulas for mixing. It is also a good idea to have a clay tool with a blunt cutting edge (models are soft!) to trim the alginate.
3. Apply petroleum jelly, and ensure that the model is comfortable.

First Piece

LAYING UP THE ALGINATE

1. Alginate has an initial setting time of approximately 2 minutes, so mix up small batches at a time (150–200g) so that it does not set before application to the model. The mix proportions are approximately 50 : 50 water to alginate, which should produce a slow-running paste. Use cold (or even iced) water to retard the setting time. Add the water to the alginate and mix vigorously and thoroughly to achieve a smooth paste.
2. Once smooth, immediately start applying to the skin with a wooden spatula. The alginate needs to be built up to a thickness of approximately 10–15mm, to a little over the halfway division line to allow for trimming. As it is applied it will start to run off the hand so you will need to keep pasting it back onto itself.
3. Push the alginate into the skin detail with the spatula to expel any trapped air.
4. Apply adjoining batches rapidly to ensure adhesion. It is useful to have an assistant.
5. Use a clean cup for each new batch to avoid wasting time cleaning cups.
6. Once the whole first half of the hand is covered and firm, trim the excess alginate back to the halfway division line with a blunt clay modelling tool.

LAYING UP THE PLASTER BANDAGE CASE

Once the alginate has set it becomes relatively firm but will remain flexible. This is advantageous in aiding removal of the mould and for the removal of castings. However, it will not retain its original shape once removed from the model, so will need to be backed up with a rigid plaster bandage case.

Apply the plaster bandage directly to the alginate (no need for release agents), following the same method as for the plaster bandage mould previously explained, thus creating a rigid supporting case for the alginate. It is important to create a flange around the seam of the plaster bandage as before, to allow for accurate registration between mould pieces.

When the case is in place, the first half of the mould is complete and the second half can be made.

Second Piece

1. Prepare all materials as for the first piece of the mould.
2. Flip the model's hand over and make them comfortable, taking care to keep the first mould piece in position on the hand.
3. The exposed side of the hand will need to be cleaned and petroleum jelly applied as for the plaster bandage mould above. Petroleum jelly will also need to be applied to the exposed flange of the plaster bandage and alginate of the first piece, so that the two halves of the mould do not stick together.
4. Follow the procedures as for the first mould piece, creating the second piece tightly up against the first.

ABOVE: **Applying petroleum jelly to the model's hand.**

ABOVE RIGHT: **Applying alginate with a plastic palette knife.**

RIGHT: **Trimming back alginate to a good edge.**

Opening the Mould and Casting

▦ Follow the opening procedures as for the plaster bandage mould above. It may be advantageous to remove the case first, then peel the alginate off and lay it back into the case.

▦ An alginate mould should be cast into immediately after removal as the alginate will start to dry out and shrink. If it is not possible to cast immediately, the mould should be wrapped in a damp cloth and sealed in a plastic bag, but should be cast within 48 hours.

▦ Before casting, the inside surface of the plaster bandage flange should be covered with petroleum jelly in case of leaks and the two halves of the mould secured together with a double strip of plaster bandage.

▦ A plaster cast is generally taken from an alginate mould, and, once set, the plaster bandage case should be removed and the alginate peeled off in pieces. As with the plaster bandage mould, one cast is usually all that can be produced, the mould being wasted during cast removal.

Life Moulding from the Face

Do not attempt to take a life mould from the face unless you are familiar with the materials and procedures involved. Familiarize yourself by moulding from a safe part of the body first. Only ever use plaster bandage or alginate. Your model will need to feel that you are in control and that they are safe; explain the procedure before and during the mouldmaking process.

OPPOSITE PAGE:

TOP LEFT: **Sqeezing out excess water from the plaster bandage for the case.**

TOP RIGHT: **Applying plaster bandage. Rub the plaster through the bandage pores in situ.**

MIDDLE LEFT: **Fold the bandage back onto itself to form a flange.**

MIDDLE RIGHT: **Applying the second part of the alginate.**

BOTTOM LEFT: **Part the set two pieces of the completed mould.**

BOTTOM RIGHT: **The mould is treated with release agents and sealed with bandage ready for casting. (See pp.121–2 for casting technique.)**

Either of the two methods described for making the mould from a hand can be used. The preparation, methods and procedures of using the materials should be followed as for the hand mould. There are, however, a few additional points to be taken into consideration.

▦ Protect the model's clothing with a sheet of polythene or cloth.

▦ Ask the model to remove any contact lenses.

▦ It is not possible to take a life mould from the hair unless it is very short and thickly covered with petroleum jelly, so a tight-fitting swimming cap should preferably be worn.

▦ Petroleum jelly should be liberally applied to any hairy bits, particularly eyebrows and lashes. If the eyelashes spring out from the jelly, reapply just before covering the eyes with the moulding material.

▦ The eyes and mouth should be kept closed at all times during the procedure.

▦ The model will need to breathe through their nose during the process so ensure that their nasal passages are clear. Wide-bore drinking straws can be placed in the nostrils to facilitate breathing. These should be approximately 80mm long and should be put in just before the nostril area is reached, otherwise they will get in the way.

Model is protected with plastic cover.

- Work from the neck up, so that the eyes are covered last.
- If attempting to mould from the ears it is advisable to plug them with cotton wool to prevent mould materials entering too deeply into the ear. Remember that with the ears blocked the model may feel even more isolated in the mould, so ensure that they are kept aware of the procedures and progress while making the mould.
- If taking a mould from the whole head, the mould will need to be created in several pieces. The lines of division should run from the highest point of the head, down the sides and around behind the ear, continuing down the sides of the neck. If using just plaster bandage another line of division should be determined in front of the ear, defining two separate mould pieces for the ears. The front of the mould will incorporate the face; the back of the mould will incorporate the back of the head and neck. The back piece of the mould should be created first, the second piece then built up tightly against the first and so on. When moulding the whole head the mould should be removed as as soon as possible so that there is still a little flexibility in the plaster bandage to bend over the ears. It is possible to create an alginate mould of the whole head in just two pieces, front and back, as long as the surface of the alginate is not undercut over the ears, allowing the plaster case at this point to be bent away from the alginate while removing (*see* diagram 2).

- When demoulding, as with the hand mould a collaboration between model and mouldmaker is essential for the comfort and safety of the model, and to prevent damage to the mould. The model can gently flex the muscles of the face to aid the demoulding procedure.
- Casts should be made hollow, as the weight of a solid cast may distort and damage the mould (*see* Chapter 10).

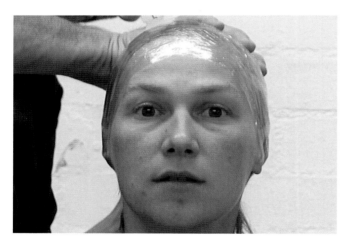

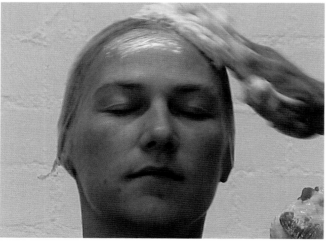

ABOVE LEFT: **Hair is masked with cling film or tight-fitting bathing cap.**

ABOVE: **Petroleum jelly is applied.**

LEFT: **Alginate is mixed and application begun.**

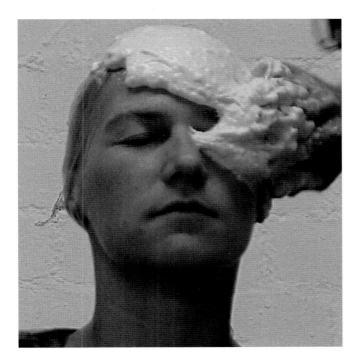

Carefully lay alginate over the face, leaving the nose exposed until last.

Finally cover the nose area, ensuring an assistant keeps the nostrils clear.

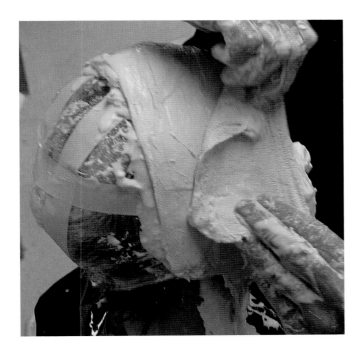

A plaster bandage case is applied over the alginate mould. The plaster bandage should be soaked and applied in multi-layered batches to speed up application.

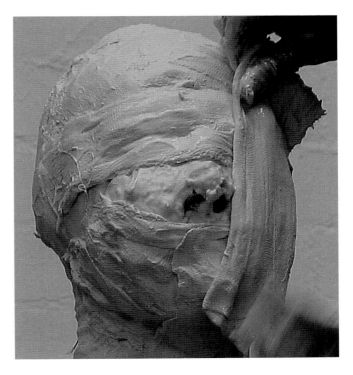

Ensure that the case does not cover the nostril area.

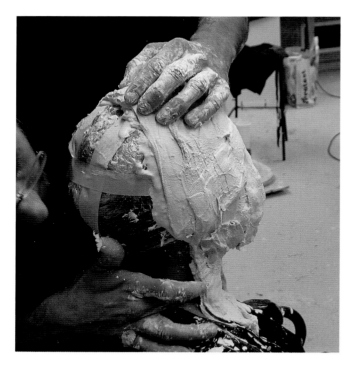

Once the case is firm the mould can be eased off the model.

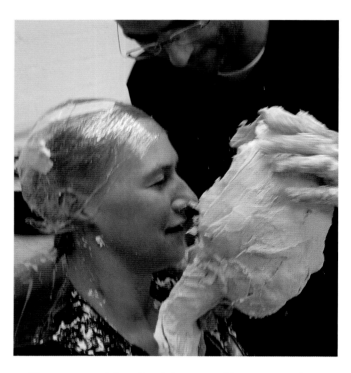

The case containing the alginate mould is removed from the model.

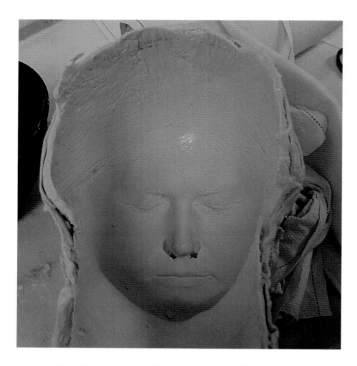

The finished mould and case ready for casting.

The mould is filled with plaster immediately to produce the cast.

Making Life Moulds from Other Parts of the Body

Follow previously described procedures for the hand, face and head moulds. Some additional points should be considered.

■ Alginate is a relatively expensive moulding material, so it may be preferable when making a whole body mould to use a combination of alginate and plaster bandage mould pieces. Alginate could be used on areas with the most important detail like the hands and face, and the other areas of the body moulded with plaster bandage. When taking moulds from the ears better results can be achieved using alginate.

■ If attempting to mould the whole body the mould will need to be made in several pieces.

■ Making a mould of the whole body is a lengthy process, so mould pieces will have to be removed when the model needs to rest. When resuming moulding, the previously made mould piece that connects directly to the next to be made will need to be put back onto the model, so that the previously made piece will connect tightly and accurately to the piece being made.

■ Light timber splints can be used as additional reinforcement; reinforcing bars should be made of a size relative to the size of the mould piece.

■ Light pieces of timber can be used where limbs are crooked. These should span across the angle of the crook and be securely plaster-bandaged in position.

■ Casts should be constructed hollow for economy of materials, weight and to prevent damage to the mould.

■ Where possible, mould pieces should be securely connected while casting. This will only be possible where there is enough access into the mould to connect the cast sections.

■ It is possible to connect cast sections of a body cast once they have been removed from the life mould, but it should be noted that a lot of post-cast finishing will be necessary.

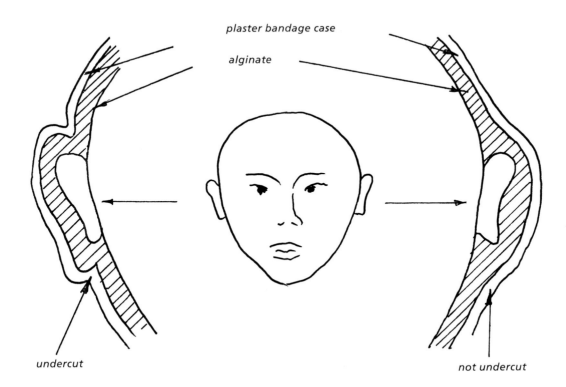

plaster bandage case

alginate

undercut

not undercut

Diagram 2

CHAPTER 10

PLASTER CASTING

The Basic Principles of Plaster Casting

This chapter deals with the basic principles of casting, that is, the production of the positive form from the negative mould. Before deciding in which material the castings from your mould are to be produced, there are some essential considerations that are common to all casting processes whichever material is being used. The decision as to which casting material to use in the mould could depend on a number of factors, including aesthetics, cost, durability, weight and so on, and these decisions can be better made with an understanding of the basic principles involved in the casting process.

Probably the most basic decision to be made is whether your cast will be solid or hollow. In theory, solid or hollow casts of any material could be produced from any mould, but in practice a number of factors should be considered.

- A solid casting will obviously use more material than a hollow one and this will have a bearing on the cost if a run (more than one) is to be produced.
- A solid casting will be heavier, which may present problems in production if using large moulds; handling and transport of the casting will also be a consideration.
- Hollow casting will use less material, lowering the weight; this may be a disadvantage if the weight of the casting is part of its aesthetic, but will reduce production costs.
- Large hollow castings may need to be reinforced with timber or steel from within to provide strength.
- Large moulds to be filled solid may have to be filled gradually to avoid the casting material cracking, which will lengthen production time.
- Hollow castings will generally need to be laminated with a reinforcing material to provide strength, lengthening production time.
- If there is restricted access through the opening into a mould made in several pieces, hollow casts will need to be made with the mould pieces open and then put together to join all the cast pieces. The edges of the cast pieces will need to be made slightly thicker so a positive join can be made between pieces. Care should be taken to avoid cast material getting on mould edges.
- Small moulds made in pieces may be too small to cast hollow. Large moulds may be too large to make solid casts practically.

Plaster

Casting plaster is a very different material to the domestic building plasters that some may be familiar with. The principal difference is the particle size of the powder, which affects the mixing, setting and hardness of the plaster. Compared to the building plasters, the particle sizes of casting plasters are very small, which gives them a much greater strength and means that these plasters will pick up very fine detail from a mould. This particle size varies between different casting plasters, producing plasters of different strengths and with differing mixing and setting properties. The differences can be categorized into 'Alpha' and 'Beta' plasters.

Alpha Plaster

Alpha plasters have a very small particle size, producing very hard plasters when set. Because of the small particle size these plasters require a higher ratio of plaster to water when mixing. When the water has evaporated from the mix, the set plaster will be left with a high percentage of plaster particles, producing a strong plaster. Alpha plasters generally have a much longer working time than Beta plasters, which may be an advantage when laying up as a hollow cast or case for a flexible mould. When used in this way, much thinner casts or cases can be achieved, economizing on material.

Beta Plaster

Beta plasters have larger particle sizes compared to Alpha plasters and therefore a relatively weaker set strength. They still have good detail productive characteristics, although not quite as good as the Alpha range. A slightly softer plaster may have other advantages, however – finishing or post-carving of the cast will be considerably easier, for instance. Beta plasters have a quicker setting time, which may also be an advantage if long runs of solid casts are required. (For the full range and properties of different casting plasters, *see* manufacturers' specifications.)

Storage and Handling

Casting plaster can be purchased in thick paper bags or sealable plastic containers. Bags or containers should be date-stamped when manufactured and should ideally be used within three months of this date. It can be kept for up to five or six months if stored sealed in a dry place at a temperature of not less than 13–15°C. Working/setting times and strength will be compromised if the plaster is allowed to absorb any moisture, so if in doubt a small test batch should be mixed before embarking on the work. Hands should be dry before opening bags or containers and while dispensing. Containers should be sealed and bags folded and closed after each mix.

Tools and Equipment

Hands and a good mixing bowl are the principal pieces of equipment required for using plaster.

Cheap plastic bowls readily available from builder's merchants are fine for this purpose, but slightly more expensive flexible rubber bowls will enable the user easily to break out any set dry plaster left over from a mix. Other principal tools are surforms and rasps for paring-back set or setting plaster. Various sizes of flat and curved surforms and rasps are useful for trimming moulds for ease of reassembly and handling, and for trimming castings prior to finer finishing.

Other useful tools include:

- chisels for chipping out casts from waste moulds. Various sizes of old wood chisels are ideal. Stone claw chisels are useful for removing a lot of material quickly;
- a selection of steel modelling tools;
- hammer;
- various sizes of steel scrapers of the sort used in the decorating trade;
- saw: it is possible to trim set plaster with an ordinary hardpoint woodsaw or general-purpose saw with an adjustable angled blade;
- knives. Heavy bladed for hacking, strong bladed with a rounded point for parting mould pieces, Stanley knife for light trimming;
- old hacksaw blades for trimming;
- plaster float and trowel for larger work;
- wire brush for clearing tools of old, set plaster;
- buckets for washing bowls and tools prior to sink washing;
- paint brushes (decorators') for application of plaster. Hard plaster brushes can be 'beaten out' and re-used.

Plaster tools. (Top) Wire brush. (Middle L–R) Surform, surform, plasterer's trowel, rasp, rasp. (Bottom) Hacksaw blade and holder, scraper, scraper.

Studio Practice, Personal Protective Equipment and Cleaning

A warm, dry place for storage and a strong sink with a plaster trap are probably the most important requirements. A plaster trap is a tank that catches excess plaster and allows it to settle before it has a chance to enter the main drain, where it would inevitably set and cause a blockage. When the tank is full it can be disconnected and the excess plaster discarded. Plaster bowls and equipment should be washed in a bucket prior to sink washing even in a sink with a plaster trap.

Measuring and Mixing

During the manufacturing process of plaster, the mineral gypsum ($CaSO_4 2H_2O$) is heated until the water part of it (H_2O) is driven off, leaving a dry powder (a process known as calcification). When the plaster is mixed with water a chemical reaction takes place in which the plaster takes back the water and sets to a solid again of similar properties as the original mineral.

There are some very precise methods of mixing plaster using weights and measures of plaster to water, but in practice most casting plasters can be mixed 'by eye', provided certain set procedures are followed (some 'high spec' alpha-range plasters, however, do require very accurate mixing; follow the manufacturer's recommendations for these).

Plaster is added to water, and the ratio of plaster added to water is largely dependent on the particle size of the plaster being used. If using an Alpha plaster the particle size will be smaller, so a larger amount of plaster will need to be added to the water to obtain the right consistency mix; conversely, Beta plaster, with a larger particle size, will need less plaster to water for a mix. For a standard workable mix the consistency should be that of double cream and should coat the hand without it being possible to see the skin. At this consistency a mix can either be poured or laminated-up, as described later. To obtain this consistency mix, the same 'by eye' procedure can be employed for both Alpha and Beta plasters.

Clean bowls and water should be used for every mix produced. Dirty mixing bowls with old dry plaster still in them can affect setting times as the dry plaster draws off liquid from the mix, and contaminated water can affect set strength and setting times. Generally, cold water is used for standard setting times, but warm water can be used to speed up setting times.

Pour the required amount of water for the batch to be made. As an approximate starting point for a batch, Alpha plasters require about two-thirds plaster to water and Beta plasters about half and half. Retarders can be used to extend the working times of casting plasters and should be used in accordance with the manufacturer's instructions. Mix batches of a size appropriate to the working time of the plaster you are using.

1. Begin by sprinkling plaster evenly over the surface of the water with one hand. It is useful to keep one hand clean and dry for other jobs (or to answer your mobile!). Continue to add plaster to the water by the handful, but do not mix; let the water absorb the plaster.
2. As the water absorbs the plaster and starts to build up under the surface, start to slow down, using smaller handfuls until the plaster is just under the surface of the water.
3. With a last handful or two, add plaster until it just starts to come above the surface of the water in small, dry islands. At this stage, there is enough plaster in the mix and it should be left for a minute or two to soak – this is important to allow the plaster to become fully saturated before mixing. A batch left like this will not start to set for a long time as it is only when mixing that the chemical reaction really starts the setting process. Once soaked for a minute or two, the surface of the batch will appear like 'watery porridge' and is ready to be mixed.
4. Mixing can be done with a tool (spoon, spatula, etc), but is much more effective by hand (rubber gloves or barrier creams can be used on sensitive skin). Mix with one hand, the fingers open and the hand kept under the surface of the batch to avoid introducing excess air to the mix.
5. Squeeze out any lumps and discard any hard bits of plaster. Agitate and stir the mix until of a smooth, homogeneous consistency; the consistency should be the same as double cream and should coat the hand without it being possible to see the skin. If the mix is too thick it will not pour or lay up evenly and if too thin it will have too little set strength. Once the batch has been mixed it is difficult to add more plaster or water if it is too thick or thin as this will produce lumps in the mix. These lumps can be squeezed out, but this takes time and effectively produces two differently setting mixes in one batch, potentially affecting working time and set strength. This may be possible with smaller batches but with larger ones it may be preferable to start a new mix.

MIXING PLASTER

1. Plaster is sprinkled by the handful evenly over the whole surface of the water.

2. Allow the water to absorb the plaster as it is added until it appears just over the surface of the water; it should have the appearance of 'watery porridge'.

3. Mix with the hand under the surface of the mix.

4. Once mixed, it should be smooth, free of lumps and should coat the hand evenly.

6. It is worth mentioning that sometimes a thicker 'paste-like' plaster may be required, for instance for reinforcing moulds or casts rapidly. If this is the case, extra plaster should be added to the water prior to mixing. Mixes should not be thinner than the standard mix, however, or the set strength will be compromised.

7. When the mix is smooth and is of the correct consistency, the batch is then ready to use within the working time of the plaster.

Colour (Integral)

Plaster can be coloured at the mixing stage to produce an integral colour throughout the plaster. For light tints, any water-based paint can be used, but for more intense colours earth pigments (available from art shops) should be used. It is not advisable to use too much colour of whatever sort as this may affect the set strength of the plaster; tests should therefore be carried out before committing to actual castings.

Principal Casting Techniques

Pouring

When making small castings, plaster can be poured to produce solid casts. In theory, any size of casting could be poured solid, but for very large castings this would be uneconomical and impractical. When plaster is setting, an exothermic (heat-generating) reaction takes place, and so the larger the volume of plaster poured, the longer it will take for this heat to dissipate, potentially cracking the casting. Plaster will also shrink to varying degrees, depending on which plaster is being used, and although this is not usually a problem with small castings it can cause problems with larger castings. Although not a cardinal

rule, I would not recommend pouring solid casts larger than a portrait head; smaller solid castings should be perfectly safe.

Ensure that the moulds are clean and dry (any pooling of water in a mould may produce marking on the surface of the cast). Moulds that have their opening at what will become the base of the cast should be temporarily secured to the workbench with the opening level; remember that plaster is a liquid and will always find a level when poured as a mass. Moulds that are in pieces and that will be filled through a pouring hole, should be temporarily secured with the hole at its uppermost position.

Plaster can now be poured to fill the mould. It should be of the standard mix consistency described in the previous section, steadily and preferably in one operation. Moulds can be topped up if not enough plaster has been mixed, but the first pour of plaster in the mould should be de-aired first.

De-airing the mould involves removing any air that may have been trapped as the plaster is poured in the mould. On small moulds this can be achieved by tapping the sides of the mould until small bubbles appear on the surface of the plaster. Larger moulds may need to be lifted and shaken gently from side to side. Another method is to agitate the surface of the plaster with the fingertips. Dip the fingertips into the surface of the wet plaster and move them up and down vigorously without taking them out of the plaster. This forces the plaster into the mould, displacing the air that rises to the surface (think Archimedes!). De-airing small moulds can also be achieved by gently blowing the surface of the poured plaster. Once filled, any overspill can be scraped back with a straight edge before the plaster sets. Piece moulds with a pouring hole should be filled to the top of the hole to allow for shrinkage and trimming.

The alginate and plaster bandage mould ready to be filled.

The mould is secured upright.

The plaster is poured steadily into the mould.

ABOVE: Fingertips are introduced into the plaster and agitated to bring air bubbles to the surface.

ABOVE RIGHT: After the cast has set fully, the plaster bandage case is removed from the alginate mould.

The alginate mould is peeled away to reveal the cast.

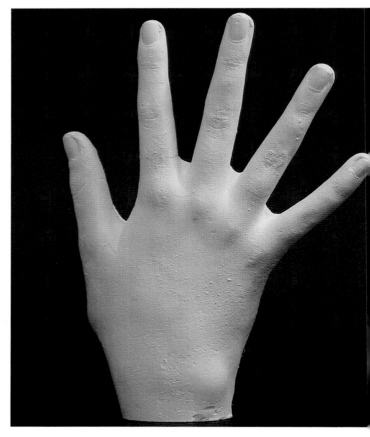

The finished cast.

Laminating

Laminating plaster in a mould is the alternative to filling it solid, using plaster built up in layers with a brush and then reinforced with a scrim (loose-weave hessian or fibreglass mat). Obviously this method is much more economical in terms of materials and the weight of the resulting cast, and is really the only option when producing very large castings, although it can also be employed for the smaller castings. This method is a little more labour-intensive, so time may be a consideration. Access into the mould is also an important consideration.

If the opening into the mould is large enough to allow access by hand to the whole interior of the mould then there is no problem. If not, then the mould will inevitably have been constructed in pieces, and plaster will therefore either have to be applied separately within those mould pieces and then reassembled to join all the cast pieces, or the mould will have to be partially assembled to allow enough access while casting and then the remaining piece or pieces joined.

It is worth mentioning here that the techniques employed for laminating plaster casts can be adopted for making cases for flexible moulds, the only difference being that you are working on a convex rather than a concave surface.

LAMINATING (EASY ACCESS)

Ensure that the mould is clean and dry (a wet mould can leave water marks on the surface of the cast), and prepare the standard mix of plaster as described previously. Laying up plaster consists of building up layers, each of which, at least initially, should be allowed to set before laying up the next. While laying up, the plaster should remain at the standard consistency; if it starts to thicken it should be discarded and a new batch made up. For this reason, try to make batches of a size that can be used within the working time of the plaster before they thicken up.

1. Begin by loading a brush with plaster and just lightly touching the surface of the mould to deposit the plaster. The brush should not be used as a paint brush, rather as a means of lifting the plaster from the bowl and depositing it onto the mould – the brush hardly needs to touch the surface of the mould to do this. If the brush is used as a paint brush the plaster will be streaked and deposited unevenly. Once the brush has deposited its load, immediately reload and lay up on the next area of the mould.

2. Plaster at this consistency and laid up in this way is 'thixotropic', that is, it will not run off a vertical surface. If the mould surface has been over-applied with water-repellent release agents, only small deposits of plaster may adhere, in which case either allow the mix to thicken and reapply or let what has adhered set and build up again with the next layer.

3. Continue by laying up brushfuls one after the other quite rapidly until the whole surface of the mould is covered. This initial layer will only be 2 or 3mm thick, but is the crucial layer that will pick up all the detail in the mould; for this reason, be particularly careful applying it, taking care to cover the whole surface evenly. Allow this layer to set before applying the next.

4. The amount of time allowed for this initial set will depend on which plaster is being used, but a general rule of thumb is that when the shine goes from the plaster it is beginning to firm up. The plaster does not need to have reached full hardness (you could still break through it easily with a finger); it just needs to be strong enough to allow a second layer to be laid on top with a brush.

5. Build up the second layer in the same way as the first. Because the first layer is slightly drier, it will drag moisture out of the second, providing a good lamination between layers. This will also speed up the setting time a little, which will enable the third layer to be laid up sooner.

6. After the first two layers have been applied the mould detail will have been well and truly covered and will be 'safe'. The cast should now be approximately 4–6mm thick, and just needs to be reinforced with a scrim to the final thickness.

7. Generally speaking, casts do not need to be more than 10–15mm thick; however, the final thickness of the cast will depend on the type of plaster being used and the size of the cast. If using a Beta casting plaster a third layer may need to be applied by brush before reinforcing with scrim. With a combination of Alpha plaster and 'fibreglass' scrim, very thin but exceptionally strong casts are possible. Whatever the thickness of the cast, a general rule of thumb is that the scrim layer should sit sandwiched within the final third of the casting.

LAMINATING THE SCRIM

It is important to use some form of scrim to reinforce a cast that is laminated, as it adds necessary integral strength. The traditional scrim material is an open weave hessian that is available in rolls approximately 50mm to 1m wide.

1. Enough strips of scrim should be cut and prepared to cover completely the layers of plaster previously laid up. The scrim needs to be run through a bowl of plaster and then applied to the cast surface, so needs to be of a length and width that can be handled by hand. If using fibreglass, 300g chopped strand should be used, and split to produce a more open weave.
2. Run strips of scrim through a bowl of standard-mix plaster until completely saturated. Lay them on the surface and pat them firmly to remove air bubbles and ensure a good lamination. It is advisable to keep scrim material a few millimetres away from the mould opening as this will be the base of the cast that may need to be trimmed.
3. Continue laying strips of scrim, overlapping by a few millimetres, until the whole surface is covered. One layer of scrim material should be enough, but on particularly large casts a second may be necessary.
4. Once the scrim layer has been applied, a last layer of plaster should be laid up to sandwich it completely. This can be done with a slightly thicker mix of plaster, and applied by brush or hand to make up the final thickness of plaster. If the inside of the cast is visually important, a wet brush can be used to smooth this final application of plaster.
5. The cast is now finished and should be allowed to set fully before demoulding.

Laminating – Piece Moulds

Moulds that are created in several pieces but do not have adequate access via the mouth of the mould or that are enclosed will need to be laminated open. This means that the individual pieces of the mould will need to be cast with plaster separately, and then the mould pieces connected to join the cast pieces within. We can see that this presents an additional job to join the cast pieces together within the mould.

The job of joining these pieces can be greatly reduced if the mould can be used with as many of its pieces together as possible, while still providing enough access by hand to apply the plaster. Inevitably, it will sometimes be necessary to laminate all the mould pieces separately and then join them.

The procedure for laminating separately is exactly the same as for other moulds, with some exceptions, notably the mould edges, where the pieces will connect when reassembled. The edges should be laminated a little thicker if possible for strength and should be 'chamfered' slightly (angled down into the mould). Chamfering of the cast edges will allow the cast pieces to be assembled accurately and will reduce seam lines.

Similarly, it is also important to keep plaster off the mould piece edges as this will impair accurate reconstruction of the mould, resulting in visible seam lines on the cast.

Joining Cast Pieces

With the cast pieces laid up, they should be allowed to firm up but preferably not dry out. It is always better to join the cast pieces within the same day, while they are firm but still damp, for two reasons. Firstly, cast pieces that are allowed to dry out completely may warp the mould pieces, making them difficult to reassemble accurately, and, secondly, the edges of cast pieces that are dry will need to be thoroughly soaked with water to enable a strong joint to be created. Edges that are slightly dry, even within a few hours of making, should be wetted anyway.

1. Reassemble the mould pieces securely, ready for a mix of plaster to be introduced into the mould to join the separate cast pieces.
2. Pour a standard mix of plaster into the mould mouth or pouring hole. Initially, this should be just enough to create a 'slush coat', a deposit of plaster a few millimetres thick.
3. The mould should now be revolved, 'slushing' the plaster around inside, covering all the cast seams. If there is excess plaster, pour it out to prevent it pooling in the bottom of the mould and setting as a lump.
4. Allow the first slush coat to set for a few minutes and repeat the process several times until certain that a good deposit of plaster has been left securely joining all the pieces of the cast. However, if it is possible to gain enough access to place scrim material to cover the cast seams this is advisable as well.
5. The cast should now be allowed to set thoroughly to its maximum strength before demoulding.

Demoulding

The techniques for demoulding, that is, removing the finished cast from the mould, will largely depend on what type of mould is being used.

■ **Plaster one-piece** Moulds can be tapped or levered out from the *inside* of the cast using a chisel. Soaking in hot soapy water can help to ease out difficult casts.
■ **Plaster multi-piece** Moulds should be eased apart piece

PLASTER LAMINATING FROM A TWO-PIECE MOULD

Applying the first layer by brush.

The first layer is applied to both mould pieces and allowed to firm up before subsequent layers are applied.

Scrim layer is applied and then laminated with more plaster until full thickness is achieved.

The mould pieces are carefully assembled.

The mould is secured with bicycle inner tube.

A mix of plaster is poured into the mould.

PLASTER LAMINATING FROM A TWO-PIECE MOULD *continued*

The mould is rotated to cover the seams with plaster.

If access is available, scrim is applied to the internal seam.

The finished cast.

by piece. Small moulds can usually be parted using a strong-bladed knife. Larger moulds may require a thin chisel to crack the seam initially until a wooden or plastic wedge can be driven in. On long seams, a series of wedges can be gradually driven in until the seams part.

■ **Plaster waste** *See* chipping out of plaster waste moulds.

■ **Fibreglass** As for plaster one- and multi-piece moulds, although there can be a certain degree of flexibility in a fibreglass mould depending on how thickly it has been made.

■ **Flexible moulds (without a case)** Demoulding from flexible moulds without a case involves a combination of flexing and peeling the mould while easing the cast out.

■ **Flexible moulds (with a case)** First remove the case (this should be carried out as for a plaster one-/multi-piece mould) and then peel the mould away from the cast. It is possible to peel the mould away from the cast with these moulds as they are thinner and more flexible than flexible moulds without a case.

■ **Press moulds (without a case)** Essentially as for flexible moulds without a case, except that the clay mould can be wasted, that is, peeled or picked away in pieces, thus destroying the mould.

■ **Press moulds (with a case)** As for flexible moulds with a case, but with the clay usually being wasted in the process of demoulding.

■ **Life moulds** Plaster bandage moulds should be soaked in warm soapy water and then broken away from the cast piece by piece. Alginate moulds with a plaster bandage case should be treated as for flexible moulds with a case.

Trimming, Cutting Back and Cleaning

Once the cast has been demoulded any excess cast and seam lines should be trimmed back.

- Initially, the extremities of any excess can be broken off carefully by hand or with a pair of pliers, but be careful not to break off too much.
- Then a number of tools can be employed depending on the size of the cast and how much plaster is to be removed (chisels, hacksaw blades, surforms, rasps). This initial trim should not be taken right back to the surface of the cast and should only be used to remove bulk excess.
- Smaller and smaller tools should be used the closer to the cast surface that you are working.
- The final cutting back to the surface of the cast should be achieved with decreasing grit sizes of abrasive paper. Fine files, including needle files, can be useful to get into difficult-to-access areas of the cast (a fine brass or wire brush is useful to clear clogged files).
- Oil-free or non-ferrous wire wools can be used, as these should not mark the cast surface; however, they should only be used on thoroughly dry casts.
- Once trimmed and cut back, casts can be rinsed under a warm running tap and allowed to dry thoroughly.

Surface Finishing

Casting plasters, whether Alpha or Beta, are tight-grained materials that will pick up very fine detail from a mould. From a very smooth-surfaced mould, a plaster cast can have a slight gloss to it that can be very attractive if simply left as it is. Surfaces can, however, be treated in a number of different ways to produce different effects.

- Standard furniture wax polishes can produce very attractive results. Either pre-stained or clear waxes can be used. Waxes will provide a protective coat to a cast and can be buffed to a high gloss or left as a gentle sheen. It is worth pointing out, though, that clear wax polishes may have a slightly yellow tinge and a neutral wax should be used if you want to retain more of the original colour of the plaster. Waxes should be applied as per the manufacturer's instructions.
- Another traditional surface finish to achieve an 'antiqued' effect is to use a very weak solution of cold tea applied to the surface with a sponge or brush, removing the excess as you go. Once allowed to dry, surfaces can be waxed or left raw.
- Iron powder rubbed into the surface of a cast will oxidize, producing a rusted iron effect.
- Graphite powder diluted in methylated spirit and applied by brush or sponge, then buffed, will produce a polished steel effect.
- If painting (oil- or water-based can be used) plaster casts, they should be treated as to the paint type being applied and appropriately primed and undercoated as necessary.
- Wood stains, fabric dyes and earth pigments should be thoroughly colour tested before application.
- There is a multitude of possible surface treatments that could be effective on plaster casts, but whatever you are experimenting with, as well as for the ones listed above, colour tests should always be carried out before commitment to the cast as a whole!
- Plaster casts can be made weatherproof with the addition of a plaster polymer, which is an acrylic material that is added at the mixing stage (see manufacturer for details). Exterior varnishes can also be used to weatherproof.

Faults and Repairs

- Small cracks or air-bubble holes can be filled by making up small batches of the plaster being used. It is sometimes difficult to colour match additional small batches, so colour tests should be carried out.
- Decorating fillers can be used to fill small cracks and holes, but should be colour matched first.
- Holes and cracks should be thoroughly soaked with water before filling with plaster or filler.
- When filling cracks and holes, overfill, allow to set and dry, then sand back.
- Large breaks can be glued together with epoxy resin glue (follow the manufacturer's instructions). Always allow broken cast pieces to dry fully before gluing together.
- If possible, back up glued repairs from behind with plaster and scrim material.

CHAPTER 11

POLYESTER RESINS AND FIBREGLASS CASTING

Although primarily an industrial material, polyester resins can be a very effective and versatile mouldmaking and casting material on a smaller scale. Polyester resins come in a variety of different forms depending on application and are a by-product of oil. They come in a liquid form of various viscosities, ranging from the consistency of runny honey to a thick paste. These liquids require two additional chemicals, an accelerator and a catalyst, to turn them from a liquid into a solid.

They can be cast into moulds as a solid, or hollow with the addition of fibreglass mat. Additions of pigments will produce casts with integral colour; additions of metal and stone powders will produce metal and stone finishes.

Polyester resin is a very durable casting material and can be used effectively with castings that require weather resistance, and used in conjunction with fibreglass to produce very lightweight casts and moulds alike.

Materials

Polyester Resins

Although a large generalization considering the huge range of polyester resins available, for our purposes three categories can be identified.

- **Gel coat resin** A thixotropic resin that is applied first to the mould surface and can be coloured.
- **Laminating/general purpose resin** A thinner resin that is used to saturate fibreglass mat when laminating the gel coat resin for strength. Can also carry metal and stone fillers to produce those finishes.
- **Clear casting/embedding resin** A clear resin used to produce solid, clear casts and translucent coloured casts, and to encapsulate objects within.

Polyester resins and their associated materials need to be used within a controlled environment due to their chemical nature, so should be thoroughly understood before application. *See* section on health and safety on p.133.

CATALYST AND ACCELERATOR

Polyester resin will eventually set to a solid if left exposed to air for weeks or possibly months, which would be unacceptable for casting purposes, therefore the addition of a cobalt napthanate accelerator and methyl ethyl ketone peroxide (MEKP – an organic peroxide) catalyst are necessary. These are added to the liquid resin by percentage to the weight of resin. Once the accelerator and catalyst have been added an exothermic (heat-generating) chemical reaction takes place, turning the liquid resin through several stages of setting time, from a jelly, known as 'gel time', through to a hard solid. Different percentages of accelerator and catalyst will vary this setting time.

Most commercially available polyester resins can be bought pre-accelerated, that is, with the correct percentage of accelerator already added, so it therefore usually only necessary to add a catalyst. If you are adding both, always add the accelerator first and mix in thoroughly before adding the catalyst.

The catalyst comes as a clear liquid that has the viscosity of a thin cooking oil. It is added to the resin as a percentage by weight of resin and it is therefore necessary to have a set of accurate weighing scales calibrated in one gram increments. At room temperature (20°C) average rates of catalyst are approximately 1–2 per cent: 1 per cent should be used for clear casting resins and 2 per cent for gel coat and laminating resins. These percentage additions can, however, be increased or decreased depending on need, and will change the pot life (working time) of the resin. A higher percentage decreases pot life and vice versa. One per cent additions have a pot life of approximately 20–25 minutes at room temperature. A word of warning – excessively high percentage additions of catalyst can cause spontaneous combustion and in extreme circumstances have explosive properties.

Ambient temperature and humidity will also affect the pot life and setting time of resins – higher temperatures will speed up setting times, lower temperatures and or damp atmospheres will decrease them. It is a good idea to do a test batch of the resin you are using in the conditions you are working in. The catalyst should be stored in a cool dark place and has a shelf life of approximately three months.

N.B. It is vital to be aware that organic peroxide catalyst is poisonous, corrosive and flammable. Bottles should always be marked with the relevant safety warnings by the manufacturer and should be used with caution and handled with care.

Polyester Resin Catalyst Table				
weight of resin (grammes)	per cent addition of catalyst in cc by weight of resin			
	1%	1.5%	2%	3%
62	0.5	0.75	1	1.5
125	1	1.5	2	3.25
250	2.25	3.25	4.25	6.5
500	4.5	6.5	8.5	13
1000	9	13	17	26

Percentage catalyst to weight of polyester resin equation for 1g of resin

2% – 1g resin = 0.017cc catalyst
1% – 1g resin = 0.009cc catalyst
Example:
2% – 253 x 0.017 = 4.3cc catalyst (round up to 4.5cc)
1% – 253 x 0.009 = 2.2cc catalyst (round up to 2.5cc)
note: cc = ml

GEL COAT RESIN

This group of resins have a paste-like consistency and are thixotropic, that is, they will not run off a vertical surface if applied to a thickness of 1–2mm. Gel coat resins are usually used as the first layer of a laminated cast within a mould. As with all polyester resins, they need an organic peroxide catalyst to make them set and are applied by brush to the mould surface. They set to form a hard scratch-, water- and weather-resistant surface to the casting. Gel coats can be coloured with specifically designed polyester resin pigments that create an integral colour throughout.

Although not generally recommended I have found that metal and stone fillers (fine mesh powders) can be used with gel coat resins to create metal and stone surfaces to casts. Because of the inherent hardness of gel coat resins this can make cutting back the surface of a cast to the metal or stone used difficult (but not impossible) and can greatly reduce the quantity of fillers required for a casting.

The gel time is generally between 15–20 minutes, catalysed at a rate of 1–2 per cent. Storage should be below 20°C and shelf life is approximately three months; test batches should be carried out if in doubt as to the condition after storage.

LAMINATING/GENERAL PURPOSE (GP) RESIN

These resins are a liquid with the consistency of runny honey and come in a variety of colours from mauve to blue. They require the addition of an organic peroxide catalyst to turn them from a liquid to a solid. It is desirable to purchase pre-accelerated GP resins; otherwise, percentage additions should be added as to the manufacturer's recommendations.

GP resins can be used to make metal- or stone-filled gel coats (the first coat to be applied to the mould surface). Additions of metal or stone powder are mixed into the resin in approximately equal parts by volume of resin, creating a thixotropic paste that should be thin enough to pick up mould detail, but thick enough not to run off the mould surface. When demoulded, the cast surface is cut back with abrasives and polished to expose the metal or stone surface.

Whether using metal- or stone-filled, coloured or pure gel coats, on their own these are brittle and need to be reinforced. GP resin is used in conjunction with fibreglass to reinforce the gel coat layer. Because these resins are relatively thin in consistency compared to gel coats they will soak into fibreglass mat and then set hard to create a very durable cast.

The alternative to reinforcing the gel coat with fibreglass mat is to fill the cast solid. This is done using an inert filler powder mixed with a GP resin. Equal parts by volume of filler and GP are mixed together and poured into the pre-gel-coated mould to produce a solid casting.

The gel time of GP resins is generally between 15–25 minutes, catalysed at a rate of 1–2 per cent. GP resins should be stored in a cool dark place and have shelf life of approximately six months; test batches should be carried out if in doubt as to their condition after storage.

CLEAR CASTING/EMBEDDING RESIN

These resins, as the name suggests, are clear and colourless and can be used to embed items in the cast. Although described as clear, different qualities and brands may have variations in clarity and colour, and so test batches should be carried out before using them.

Clear resins cannot be cast hollow, that is, built up in layers within a mould to create a hollow casting, and should always be cast solid in blocks. For this reason catalysing clear resins is crucial. All resins generate an exothermic reaction (heat generation) when catalysed, but when catalysing clear resins the exothermic reaction becomes more intense because of the greater volume being used. For this reason, when catalysing clear resin low percentage rates of catalyst should be used to reduce the exothermic reaction. Increased percentages can generate cracking and/or yellowing of castings as they set.

When embedding items, very clean working conditions are vital in order to avoid dust contamination. To embed items an amount of resin is first poured into the mould and allowed to set, the item is then introduced and covered with more resin, suspending the item within the block. Certain restrictions as to size and the nature of the items to be embedded should be observed (see 'Principal Casting Techniques' below). Clear resins can be coloured with specific translucent polyester resin pigments that will produce an integral translucent colour.

The gel time of clear resin can be between 25–35 minutes, catalysed at a rate of 1 per cent, but it will vary greatly depending on the volume of resin being used and the ambient conditions. Overall setting times are much more gradual than with other resins and should not be increased with the addition of more catalyst because of the exothermic reaction. Storage should be between 5–25°C in air- and light-tight containers. The shelf life is between three and six months and test batches should be made up after this period to check on their condition.

Laminates

Laminates generally come in the form of various grades of fibreglass and are used as reinforcement to the relatively brittle gel coat that is initially laid up within the mould. There are many grades of fibreglass that can be used for laminating different jobs, but they can be broken down into four groups.

CHOPPED STRAND MAT

This is made up of fibreglass strands between 10–50mm long bonded together with an emulsion. When saturated with a GP or laminating resin this emulsion melts, softening the mat and enabling it to be pushed into the detail of a previously laid-up gel coat. Once set, this provides great strength to a casting.

SURFACE TISSUE

This is a much finer emulsion-bonded fibreglass mat that is used in the same way as chopped strand mat, in conjunction with a GP or laminating resin, as the heavier version. Being much finer, surface tissue can be useful as a back-up to chopped strand mat if a smooth surface is required. It can also be used on its own if semi-translucent casts are required, but should be laminated to several thicknesses to provide strength.

CHOPPED STRANDS

This consists of 5mm loose fibreglass strands. Chopped strands can be mixed with a GP or laminating resin to form a paste that can be very useful as a laminate in moulds that have areas that are difficult to access. It can be used as a laminate on its own or in conjunction with others.

WOVEN LAMINATES

These are randomly woven strands of fibreglass, usually made up into a cloth or a ribbon, that are used where great strength and impact resistance are necessary. Because of the tightly woven nature of fibreglass cloths and ribbons, compared to chopped strand mats, they should be very carefully laminated as they will not soak up the resin as readily. They can be used in conjunction with other laminating materials in areas where greater strength is required. Storage of fibreglass laminates should be in a cool dry place.

Fillers

Fillers can collectively be described as very fine mesh powders that are mixed into a GP resin. They can be broken down into two main groups.

Inert Fillers

These are very fine inert mineral powders that are mixed in equal parts with a GP resin and poured into a pre-gel-coated mould, creating both the reinforcement for the gel coat and a solid casting. Mixing the resin with a filler reduces the amount of resin needed to fill a cast solid, which is useful for two reasons: to lower the exothermic reaction and to lower the cost.

Calcium carbonate can be added to GP resin to produce a very white casting. Inert fillers can also be used to add to gel coats to increase the surface strength and scratch resistance. Storage should be at moderate temperatures in dry conditions.

Metal Fillers

These are fine mesh metal powders added to resins to produce castings with metallic finishes. Unlike metallic colours these are powders of the actual metal and once cut back with abrasives and polished will appear and react as the actual metals – for instance, an iron-filled casting will start to rust if exposed to moisture.

Metal fillers can be added to resins to produce gel coats painted into a mould and then backed up with fibreglass or poured to produce solid castings. It should be considered that when creating solid metal-filled castings only the surface of the casting will have any metal visible, the rest of the metal being internal. This may be prohibitive in terms of cost. Storage should be at moderate temperatures in dry conditions. (Stone powders can also be added to resins as above.)

Tools and Equipment

Following are lists of the tools and equipment needed for the various processes involved in the casting of polyester resins and fibreglass.

Dispensing and Catalysing

- **Waxed paper cups or white plastic containers** For dispensing and mixing batches of resin. Plastic containers can be allowed to dry after use and the set resin cracked out to be reused.
- **Disposable wooden spatulas** For mixing catalyst, pigments and fillers into batches.
- **Catalyst Dispensing Bottle** Calibrated in millilitres, this is an invaluable, safe and accurate piece of equipment for dispensing catalyst.
- **Scales** Accurate scales are essential to calculate safely weights and measures of resin and catalyst.

Laying-Up

- **Brushes** Brushes of various sizes are needed to paint gel coats onto mould surfaces and to saturate fibreglass mats when laminating. Specially manufactured brushes are available that will withstand the harsh chemicals involved and not lose bristles; however, ordinary decorating brushes of a good quality will last quite a long time.
- **Scissors** For cutting laminates into appropriate sizes for the job required and trimming castings before they set too hard.
- **Disposable wooden spatulas** For mixing and applying batches of chopped strand laminates.
- **Plastic paint kettle** For cleaning brushes and equipment in acetone.

Trimming and Finishing

- **Rasps and files** Can be used for initial trimming of cured resin casts.
- **Hacksaw** Used for initial trimming.
- **Tin snips and heavy duty shears** Used for initial trimming.
- **Powered jigsaws and multi-tools** Can be very useful to take some of the hard work out of trimming (observe strict dust control measures).
- **Power drill** Holes can be safely drilled mechanically or by hand into set resins.
- **Powered abrasive tools** Useful to remove sharp set fibreglass strands. Beware of cutting back metal-filled

casts mechanically as they tend to take too much away too quickly (observe strict dust control measures).

- **Wet and dry abrasive paper** Used for removing sharp edges after initial trimming and cutting back metal-filled casts prior to polishing (use wet).
- **Needle files** For very fine and difficult to access trimming and cutting back.
- **Wire brushes** For clearing tools clogged with material.
- **Polishing equipment** For polishing castings after cutting back with abrasives. Can be mechanical or by hand.

Health and Safety – Studio Practice, Personal Protective Equipment and Cleaning

- The main studio concern when using polyester resins is adequate ventilation. All polyester resins and resin-based products emit styrene vapour, which is toxic and if allowed to build up and is inhaled can cause dizziness and nausea. Ideally, dedicated local extraction should be used in the work area, but if not available plenty of ventilation should be employed to clear potentially hazardous build-ups of fumes. Windows at either end of the working area should be opened and because resin fumes are heavier than air open doors will allow fumes to escape at ground level.
- In addition to good ventilation, personal respiratory equipment must be used. It is essential that respiratory equipment is specified for use with polyester resin – always check the manufacturer's specifications.
- Work surfaces can be protected with a heavy gauge polythene that can be disposed of or replaced when necessary.
- Overalls should be worn as polyester resins are extremely difficult to remove from clothing.
- Working with polyester resins and their associated products can cause skin reactions, particularly if you have a sensitive skin type. Barrier creams provide a protective barrier between you and these materials and should always be employed, even by those without sensitive skin. Disposable plastic or rubber gloves will provide further levels of hand protection. To clean hands, specific proprietary hand cleaners are the only efficient and safe method.

- Acetone is used to clean brushes and tools prior to detergent washing (avoid acetone contact with skin – *never wash hands in acetone!*).
- Disposal of excess resins, empty storage and dispensing containers, solvents and any other associated materials must be carried out with regard and understanding to flammability and pollutive consequences. If possible, follow the guidelines set out by the manufacturers, which should be available through the retailer.
- 'Safety Data Sheets' for all materials should be requested from the retailer if not supplied with the products. All Health & Safety labelling on materials should be understood and adhered to.

Principal Casting Techniques

Coloured, Metal or Stone-Surfaced Castings

Casting from a mould with polyester resins is generally a two-stage process whereby a gel coat resin containing colour, metal or stone filler is mixed, applied to the inside surface of the mould and then reinforced, either with a fibreglass or backfilled solid with a filler and resin mix.

Polyester resins can be cast in most types of moulds, providing that they can be released (*see* Chapter 3); the application techniques are standard to all.

COLOURED GEL COATS

To obtain casting with integral colour, pigments can be added to the first gel coat layer of the casting. Specific polyester resin pigments have been developed that are raw earth pigments suspended in a resin binder. A gel coat resin is used to carry the pigment, as it is thixotropic and can be applied to the surface of a mould without running off.

1. Dispense the amount of resin required to apply a layer approximately 2–3mm thick to the mould surface, and weigh it to calculate the amount of catalyst required (*see* the catalyst chart above). Do not add the catalyst until the resin has been pigmented. If laying up large moulds be aware of the amount of resin that can be applied within the working time or pot life of a batch. If it is necessary to lay up the

mould with several batches of gel coat, the total quantity will need to be pigmented in order to obtain colour matching, and then decanted off and catalysed in batches that can be applied within the pot life.

2. Add the pigment in small amounts until the colour required is obtained; only small amounts of pigment are needed to produce very vivid colours. Additions of approximately 2–5 per cent by weight of resin should be adequate, but in practice it is much easier to mix by eye.

3. Once the required colour has been obtained, the batch can be catalysed. The rate of catalyst for gel coat application at room temperature should be 2 per cent, which will give a working time or pot life of approximately 15 minutes.

4. Once catalysed, the gel coat can be applied to the mould surface by brush. A layer of between 2–3mm should be evenly applied and left to cure.

5. It is sometimes a good idea to apply a second gel coat to moulds with a lot of detail. In this case, allow the first gel coat to cure and dry before application of the second.

6. A gel coat catalysed at 2 per cent will be ready to laminate or backfill (fill solid) within an hour. Within an hour the gel coat will harden but remain slightly tacky to the touch, which will aid application of the fibreglass laminate.

METAL- AND STONE-FILLED GEL COATS

Metal and stone effects castings can be obtained using very fine mesh (finely ground) metal and stone powders (known as fillers) mixed into the resin.

A wide range of fillers is available including:

- bronze;
- copper;
- iron;
- aluminium;
- brass;
- marble;
- slate;
- stone.

Fillers are added to a GP resin to form either a brushable gel coat that is applied to the mould surface and then backed up with reinforcement, or a pourable mix that fills the mould. When calculating the amount of filler a rough rule of thumb is equal quantities by volume of resin to filler.

1. Dispense the required amount of resin into a container and then dispense an equal quantity by volume of the filler required. Do not mix resin and filler at this stage as the quantity of catalyst for the resin will need to be calculated.

2. For a gel-coat mix, the quantity of resin and filler combined will need to cover the entire surface of the mould to a thickness of approximately 2–3mm, depending on the amount of detail. Moulds containing a combination of raised and deep fine detail may need a second application of gel coat once the first has set.

3. For a one-operation solid casting the combined resin and filler mix will need to fill the mould completely and be of a pourable consistency. It is not desirable to fill very large moulds solid due to exothermic (heat) reaction generated during the setting process, which could potentially crack the cast. In this case, a gel coat mix should be applied followed by a laminate. It is also much more economical to use a gel coat because it contains much less filler.

4. Laying up a large mould will require several batches of gel coat in order to be able to apply it within the pot life (working time). Colour matching of metal and stone fillers cannot be guaranteed, so if the mould will require several batches of gel coat the entire quantity needed should be mixed and then decanted into batches that can be used within their pot life.

5. Mix consistencies will depend on use – gel coats should be brushable but thixotropic, and pourable mixes need to be thin enough to run into the mould detail. Equal quantities of resin and filler are only a rough guide, and where possible more filler should be used without compromising the required consistency. Less than 50 per cent filler will not produce good metal and stone effects.

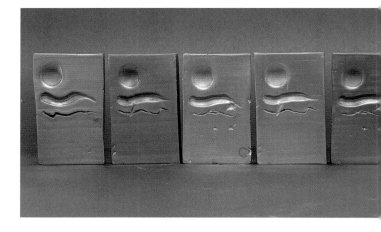

Cold-cast metal samples.

6. Once the quantities have been determined, and the resin and filler dispensed separately, the resin amount will need to be weighed to determine the catalyst quantity. Metal- or stone-filled gel coats applied at room temperature can be catalysed at 2–4 per cent (aluminium should be catalysed at 1–2 per cent). It is important that the resin and filler are not mixed together before calculating catalyst quantities as it is just the amount of resin in the mix that needs to be catalysed.

7. Once the catalyst quantity has been calculated, the resin and filler should be mixed *before* adding catalyst. When combining, add the filler to the resin a little at a time to avoid dust clouds. Metal powders can be dangerous if inhaled, and aluminium can be explosive if allowed to build up in clouds (personal protective equipment should be used; *see* above section on health and safety). Ensure that all the filler is thoroughly mixed into the resin, and that the mix has an even consistency.

8. Add the correct amount of catalyst and mix in thoroughly.

9. If a solid cast is required, ensure that the mould is level before filling it.

10. If hollow, brush the gel coat mix evenly and methodically onto the entire surface of the mould, to approximately 2–3mm thick. If a secondary gel coat is required, allow the first to set (within an hour at 2 per cent catalyst).

11. Carry out secondary lamination or backfilling.

Laminating with Fibreglass (Laying-Up)

A gel coat, whether coloured, metal- or stone-filled or plain, is brittle when set and will require reinforcement. Fibreglass impregnated with resin is one method by which to do this.

FIBREGLASS MAT

For most gel coats one layer of 300g fibreglass mat will provide a strong but lightweight casting. For very large moulds either more layers, or a heavier fibreglass, can be applied for greater strength. The ratio of mat to resin is approximately 1:2. 300mm^2 uses approximately 50g of GP/laminating resin.

1. Fibreglass mat can be easily cut with scissors, and enough pieces should be prepared to cover all the gel-coated area inside the mould. The size and shape of the pieces will depend on the shape and size of the mould – large pieces of mat can be applied to moulds with large flat areas, smaller, more detailed moulds may need smaller pieces cut to shape.

2. Lay out all the cut mat pieces in the sequence they are to be laid up on the workbench next to the mould.

3. Dispense a batch of GP/laminating resin in a waxed cup or plastic container. The laminating process involves soaking the fibreglass with resin as it is applied to the gel coat layer, so only dispense enough resin that can be used within its pot life. Start with a small cupful until you find your working speed. At room temperature resin catalysed at a 2 per cent rate is suitable for laminating. However, this rate can be raised or lowered according to working speed and atmospheric conditions.

4. Fibreglass mats have an emulsion bond that holds the strands of fibreglass together. Once the mat has been saturated with catalysed resin, the bond melts and can be stippled into the detail of the mould. Apply catalysed resin by brush to the first area of the gel coat to be laminated and then position a piece of mat on top.

5. Apply more resin to the mat until it is completely saturated.

6. Continue to the next area to be covered while the resin soaks into the first. Overlap the second piece of mat with the first by a few millimetres and saturate with more resin.

7. With the second piece of mat saturated, the first can be stippled into the mould detail. The mat should be soft enough now to be stippled with the end of the brush bristles onto the gel coat. Methodically work over the surface of the mat, pushing out any air trapped between the gel coat and mat. The aim is to achieve good lamination between gel coat and mat without trapping any air between.

8. Continue in this way, working methodically until the whole gel-coated surface has been laminated.

CHOPPED STRAND

Even when fibreglass mat softens with resin there may still be areas of a mould where it would be difficult to ensure good lamination (where there are deep hollows or very fine detail, for instance). In these cases, chopped strand fibreglass can be used to laminate. Chopped fibreglass strands are short cut strands of fibreglass that can be mixed with catalysed resin to form a paste-like mix. Use a wooden spatula to mix the strands into the resin and to apply the mix to the gel coat (brushes tend to get too sticky). Push the strands into the detailed area, ensuring good lamination and pushing out any air bubbles. Additional mat can be laminated over areas that have been covered with chopped strand if desirable.

ADDITIONAL FIBREGLASS LAMINATES

Where additional strength is required another layer of fibreglass can be applied to the first, which can be applied either straight after the first, or after it has set and dried. Where a smooth surface is required, an additional layer of surface tissue fibreglass can be applied to the heavier mat.

Areas that require very great strength can be further laminated with fibreglass roving. This consists of woven fibreglass and usually comes as a cloth or a ribbon. It should be carefully laminated as it does not soak up the resin as readily as fibreglass mat.

LAMINATED HOLLOW CAST

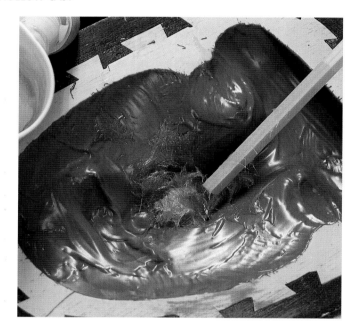

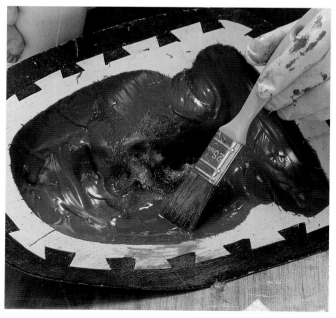

ABOVE LEFT: **Applying a bronze-filled gel coat.**

ABOVE: **Laminating with chopped strand in an area that is difficult to access by mat.**

LEFT: **A little laminating resin is applied prior to the fibreglass mat.**

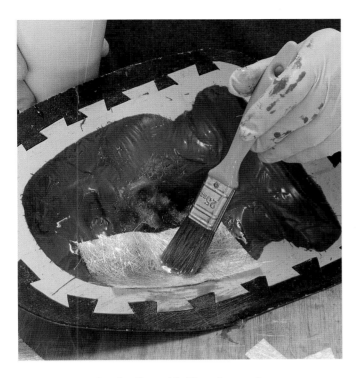

Laminating with fibreglass mat.

Once the mat is in place, it is then saturated with resin.

Air bubbles are stippled out with the end of the brush to achieve good lamination.

Excess mat can be trimmed with scissors before the cast sets.

LAMINATED HOLLOW CAST *continued*

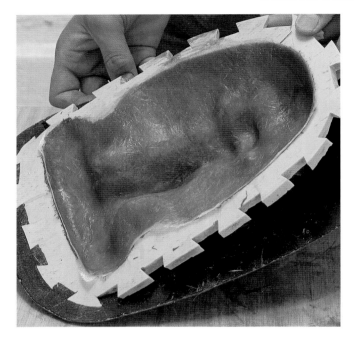

The silcone mould and cast are separated from the case.

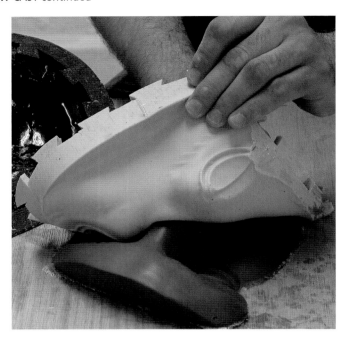

The silicone mould is peeled away from the cast.

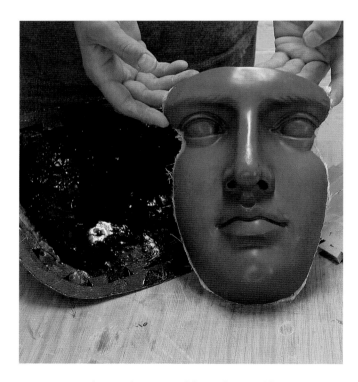

The cast is removed from the mould.

The finished cast.

Backfilling/Solid Lamination

An alternative to lamination with a fibreglass mat is solid filling with a filler powder and resin mix. It is important to use a filler powder and not just to fill the mould with neat resin. Polyester resins produce an exothermic reaction when catalysed and setting, and when poured solid in bulk may crack, so the addition of a filler reduces the amount of resin needed to fill a mould and will dissipate some of the heat generated during the curing process. Almost any finely ground powder could be used as a filler for this job as long as it does not have an adverse reaction with the resin (it must be dry). There are, however, commercially available inert mineral filler powders that have been specifically designed for the job. If weight to the cast is required, iron filler can be used as a filler. Talcum powder (unperfumed talc can be obtained from mouldmaking and casting suppliers but any type will do) can be used as a filler to back up a gel coat (mixed with less resin talcum powder can be used to produce a thick paste, which can be used as a 'car body' filler and sanded back if necessary). A mix of GP or a laminating resin can be used for this job and is mixed in equal parts by volume with the filler.

1. The resin should be dispensed first and weighed to calculate the amount of catalyst required.
2. The filler is then dispensed in equal volume, and added to the resin and mixed in thoroughly. The filler is very fine and at first will appear not to mix in with the resin, but it will eventually mix in to produce a thick but pourable liquid. If the mix is very runny a little more filler should be added a bit at a time until a mix that is as thick as possible but still pourable is produced.
3. Pour the mix into the mould. A little agitation of the mould will bring up any air bubbles and allow the mix to seep into the mould detail.
4. Allow the cast to set thoroughly and cool before demoulding.

Clear Casting

- There are many brands of clear resins available, but they vary in quality and end results. Buy the best quality available and batch-test before use.
- They vary in colour in their uncatalysed state (from water white to blue), becoming clear when catalyst is added.
- Clear casting resins have the consistency of GP resins (free-flowing viscous liquids) and are designed to be cast in blocks and cannot be built up in layers to form a hollow cast.
- Because of the exothermic reaction when resin is catalysed, particular care should be taken when catalysing clear resins. Rates of between 0.5 and 3 per cent by weight of resin should be used depending on the volume and thickness of the block to be cast. A general rule is that smaller volumes require more catalyst and larger ones less. This is because smaller volumes produce less exothermic reaction, needing more catalyst to make them set and larger volumes vice versa. Being cast in blocks, clear casting resins will always produce more heat than other resins, which potentially poses problems of shrinkage (up to approximately 3 per cent depending on the job) and/or cracking of the cast. Consideration of catalytic rates is the best way to control this. If the exothermic reaction is allowed to become excessive, it can also cause yellowing of the resin; again, control of the catalyst is the best way to reduce this (blocks of more than 500g poured in one may cause problems).
- Large moulds may require water cooling to reduce the effects of the exothermic reaction. Moulds should be suspended in a water bath to allow the water to cool the whole surface of a mould. Alternatively, sink cooling of moulds can be very efficient. A length of tube substituting the plug allows water to be constantly run into the sink and to drain out once it has reached the top of the tube, providing constantly cool water to surround the mould. Be very careful not to allow any water to come into contact with the setting resin as this will produce cloudy and tacky castings.
- Another way to reduce the exothermic effects is to pour a casting in layers. Pour a first batch of resin and cover the mould mouth with a sheet of paper to prevent dust entering. Allow this batch to set fully and cool before pouring the next batch. In this way, large blocks of resin can be produced. A word of caution here – where possible, the same amounts of resin and catalyst should be poured for each layer to avoid differences in colour between layers. Allow approximately three hours between layers to allow the resin to set and cool.

POURING A SOLID CLEAR RESIN CAST

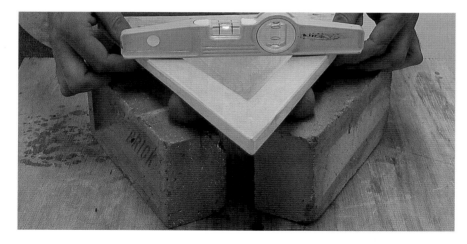

Levelling the mould. The mould is pushed down on to balls of clay until level.

Clear resin is poured into the mould.

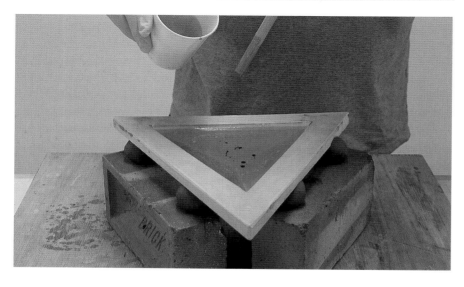

Pigment is drizzled into the resin, but not stirred too vigorously to give streaks of colour to the cast.

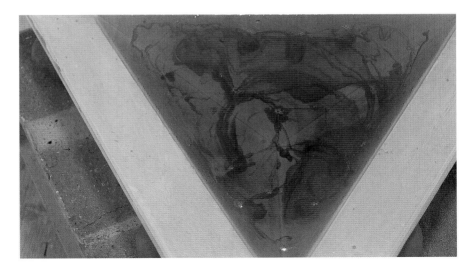

Pigment streaks in the resin.

When fully set, the cast can be removed from the mould.

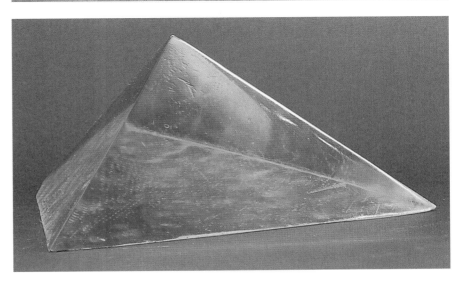

The finished cast.

POURING A SOLID CLEAR RESIN CAST INTO A MULTI-PIECE MOULD

The mould pieces are secured with bicycle inner tube rubber bands.

The mould is secured upright and resin is poured in.

The mould is rocked gently to bring up any air bubbles.

Once allowed to set, the mould pieces are separated.

The silicone mould is peeled away from the cast.

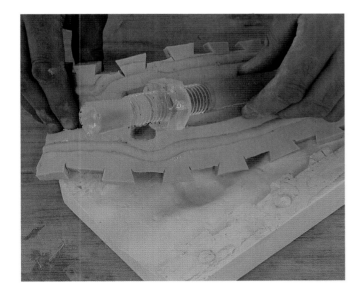

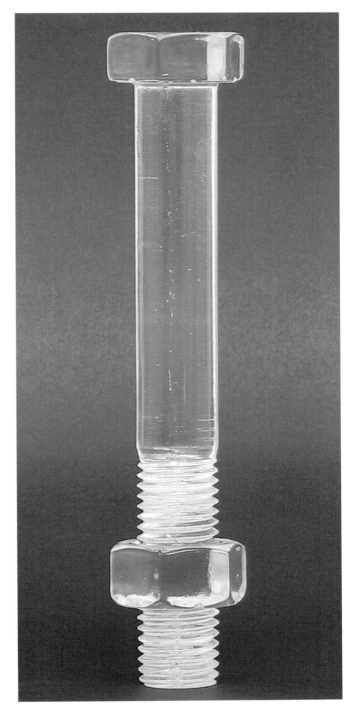

The finished cast.

Embedding Objects in Clear Resin

It is possible to embed objects in blocks of clear resin. This presents several potential problems that need to be considered if good results are to be achieved.

The first consideration is the nature of the object to be embedded. Organic materials such as plants containing water may have to be dehydrated before embedding as the resin may not cure properly if the water content is too high. Plants can be dehydrated by using a mixture of silica gel and table salt. Salt is absorbed by osmosis into the cells of the plant, taking on the water content. Silica gel is blue in colour and will turn pink when it has absorbed water. It can be reclaimed and used again if carefully dried out, whereupon it will turn blue again. Plant life should be covered in a mix of silica gel and salt in equal parts in a sealed container and left to dehydrate for seven to ten days.

Objects that are hard or have sharp edges may present problems due to the shrinkage of the resin as it cures, which can be as much as 3 per cent of the volume. Objects that are too large may crack the resin and objects with sharp edges may produce very fine lines within the block. The degree of these problems will depend on the size and shape of the object to be embedded and the amount of resin to be poured. Trial and error or test batches are the only measure of prevention.

The object to be embedded determined, the next consideration is the weight of the object. Objects that are light will float to the surface and objects that are heavy will sink. The way around this is to cast the block in two layers at the point where you want the object suspended within the block.

1. Pour embedding resin (catalysed at a rate suitable for its quantity and the volume that is to be poured) into the mould to the point where the object is to be suspended. Allow the resin to cure.
2. If embedding a light object that will float, this first pour should be allowed to cure just long enough so that the object can be stuck onto the surface preventing it from floating to the surface of the next pour. If the object is heavy, the cure needs to be just long enough to support it to prevent it sinking.
3. Place a sheet of paper or card over the mould opening while the resin is curing to prevent dust settling on the surface, particularly between layers.
4. Once this first layer has been allowed to cure and the objects placed, pour the second batch of resin. Allow the cast to cure thoroughly and cool down before demoulding.

5. If a coloured base layer is required, this should be poured into the mould first to the required thickness and allowed to cure thoroughly before proceeding to pour the clear part of the casting. GP resin can be used for this layer, coloured with polyester pigment. Clear casting resins can also be coloured with translucent polyester resin pigments that will give clarity and colour to a casting.

Surface Tackiness and Undercure

A common problem with clear casting resins is tackiness on the surface of the cast that is exposed to the air, which can be due oxygen inhibition. Oxygen inhibition is caused by styrene in the resin evaporating when in contact with the air. Where the cast has contact with the mould surface the resin has no air contact, resulting in a good cure; where it is exposed to the air styrene evaporates, resulting in undercure, when the resin does not fully cure. This is normally not a problem in other resin casting procedures where the catalytic rates are higher. When catalytic rates are reduced, as when using clear casting resins, the exothermic reaction can be reduced, causing undercure and surface tackiness. Wax in Styrene can be added to resins at 2 per cent to reduce this problem. The wax rises to the surface of the cast, thus reducing air contact and allowing a full cure. It is advisable to do test batches when using Wax in Styrene with clear resins as it may cause cloudiness.

It may not be possible to increase catalytic rates without the danger of cracking the cast, in which case any surface tackiness must be removed manually once the cast has been demoulded. In this case, immerse casts in detergent and water for several hours and then scrape off as much of the residual resin as possible. Cut back any remaining tackiness with wet and dry paper.

Aggregates

Clear casting resin can be used as a carrier for aggregates. Aggregates can be any material that will not inhibit the cure of the resin (for example, glass chips, stones, ball bearings). The aggregate material is packed into the mould, filling it completely. Resin is then poured in, filling in the gaps between the aggregate and bonding them all together (*see* Chapter 14 for alternative casting ideas).

Laying-Up Piece Moulds

Moulds that have been created in several pieces but do not have adequate access via the mouth of the mould, or are enclosed, will need to be laid up open. That is, the individual pieces of the mould will need to be cast separately, and then the mould pieces connected to join the cast pieces within. We can see that this presents an additional job to join the cast pieces together within the mould.

- The job of joining these pieces can be greatly reduced if the mould can be used with as many of its pieces together as possible, while still providing enough access by hand to lay up the gel coat and laminate. Inevitably, it will sometimes be necessary to lay up all the mould pieces separately and then join them.
- The procedure for laying-up separately is exactly the same as for other moulds, with some exceptions, notably the mould edges where the pieces will connect when reassembled. The edges should be laid up a little thicker if possible for strength and should be chamfered slightly (angled down away from the mould edge). Chamfering of the cast edges will allow the cast pieces to be assembled accurately and will reduce seam lines.
- It is also important to keep resin off the mould piece edges, as this will impair accurate reconstruction of the mould, resulting in visible seam lines on the cast.
- If possible, laminates should be laid approximately 10–20mm (depending on mould size) over the mould piece edges in order to locate into other pieces. Where laminates overlap mould piece edges, the laminate should be left free of resin and saturated just prior to joining the mould pieces.

Joining the Cast Pieces

With the cast pieces laid up, they should be allowed to firm up but preferably not dry out. It is always better to join the cast pieces within the same day while firm but still slightly soft ('cheesy') for two reasons. Firstly, cast pieces that are allowed to dry out completely may warp the mould pieces, making them difficult to reassemble accurately, and, secondly, the edges of the cast pieces that are cheesy will meld together, providing a stronger joint.

1. Whether edges are cheesy or dry they should be wetted with additional GP resin. Overlapping laminates should be thoroughly wetted before reassembly of the mould pieces.
2. Reassemble the mould pieces securely ready for a batch of catalysed resin to be introduced into the mould to join the separate cast pieces. A standard 2 per cent catalysed batch of GP resin should be poured into the mould mouth or pouring hole. Initially, this should be just enough to create a slush coat, a thin deposit of resin to provide an initial bond between the cast pieces.
3. Revolve the mould, slushing the resin around inside to cover all the cast seams. If there is excess, the resin should be poured out to avoid pooling in the bottom of the mould and setting as a lump.
4. Allow the first slush coat to gel (approximately 30 minutes) and repeat the process several times until certain that a good deposit of resin has been left, securely joining all the pieces of the cast.
5. If there is any way that access can be obtained to cover the cast seams with laminate material, this is advisable as well.
6. Allow the cast to set thoroughly to its maximum strength before demoulding.

Demoulding

The techniques for demoulding, that is, removing the finished cast from the mould, will largely depend on what type of mould is being used.

- **Plaster one-piece** Moulds can be tapped or levered out from the *inside* of the cast using a chisel. Soaking in hot soapy water can help to ease out difficult casts.
- **Plaster multi-piece** Moulds should be eased apart piece by piece. Small moulds can usually be parted using a strong-bladed knife. Larger moulds may require a thin chisel to crack the seam initially until a wooden or plastic wedge can be driven in. On particularly long seams, a series of wedges can be gradually driven in until the seams part.
- **Plaster waste** *See* chipping out of plaster waste moulds, p.35.
- **Fibreglass** As above, although there can be a certain degree of flexibility in a fibreglass mould depending on how thickly it has been made.
- **Flexible moulds (without a case)** Demoulding from flexible moulds without a case is a combination of

flexing and peeling the mould while easing the cast out.

- **Flexible moulds (with case)** First remove the case (this should be carried out as for a plaster one-piece mould), then peel the mould away from the cast. It is possible to peel the mould away from the cast as these moulds are thinner and more flexible than flexible moulds without a case.
- **Gelflex moulds** Resin casts from Gelflex moulds should be allowed to cool down thoroughly after the setting process, as the mould may soften slightly due to the high exothermic reaction and will need time to reset.
- **Press moulds (without a case)** Essentially as for flexible moulds without a case, except that the clay mould can be wasted.
- **Press moulds (with case)** As for flexible moulds with a case, the clay usually being wasted in the process of demoulding.
- **Life moulds (plaster bandage)** Moulds should be soaked in warm soapy water and then broken away from the cast piece by piece. Alginate moulds with a plaster bandage case should be treated as for flexible moulds with a case.
- **Life moulds (alginate with plaster bandage case)** The case should be removed as for plaster one-piece moulds and then the alginate peeled and picked away from the cast.

Trimming and Cutting Back

Once the cast has been demoulded, any excess cast and seam lines (flashing) should be trimmed back.

- Initially, the extremities of any excess can be broken off carefully by hand or a pair of pliers, but be careful not to break off too much. Then a number of tools can be employed depending on the size of the cast and how much resin is to be removed (chisels, hacksaw blades, surforms, rasps).
- The initial trim should not be taken right back to the surface of the cast and should only be used to remove bulk excess.
- Smaller and smaller tools should be used the closer to the cast surface you are working.
- Final cutting back to the surface of the cast should be achieved with decreasing grit sizes of abrasive paper

(wet and dry paper, used wet). Fine files, including needle files, can be useful to get into difficult to access areas of the cast (a fine brass or wire brush is useful to clear clogged files).

- Once excess flashing has been taken back to the surface, final cutting back and polishing of the entire cast surface can be tackled.

Surface Finishing and Polishing

Casts that have coloured gel coats and clear casting resins will usually come out of moulds with a matt surface finish. Metal- or stone-filled gel coats will be matt and the base material not fully visible. To obtain the full effect of metal and stone, and to achieve gloss colours or glass-clear surfaces, castings will need to be cut back with abrasive papers and then polished.

Cutting Back with Abrasives

Wet and dry abrasive papers should be used to cut back cast surfaces. They should be used wet to avoid creating dust (which is carcinogenic) and to prevent the paper from clogging. Different grit-sized papers (gauged in number of particles of grit per square inch of paper) should be used in sequence from coarse to fine. Depending on the job, 100–600 grit paper (100 is the coarsest, 600 the finest) should be used in succession, with the very fine grades usually being reserved for clear casting resins.

1. Cut up standard sheets of paper into pieces that are suitable for the size of the job and lay them out in sequence.
2. Initially use the coarsest grade just to take off the worst of any seam lines and to give the whole surface bite. With the metal and stone surfaces this will just remove the surface layer of resin, exposing the base material. Where there is a smooth area or areas that have little detail, coarse papers should be used sparingly and it may be possible to start with a finer paper.
3. As the paper becomes clogged either rinse it under a running tap or in large bowl of water.
4. Once the whole surface has been cut back with one grit size, wash the cast, change the rinsing water and move on to the next grit size. In this way, the whole surface of the cast is cut back progressively using finer and finer paper, scratches created by the coarser grades being removed by the finer grades. You will notice that the metals will become more

and more visible as the abrasive exposes them. Clear resins will remain cloudy even after the finest grade and will not become clear until polished.

5. It is difficult to gauge if casts have been abraded enough until after polishing, but thorough, methodical abrading will produce better results. If polished results are not good it usually means returning to abrasives before polishing again.

Polishing

Polishing can be done either mechanically or by hand using specifically designed polishing compounds. Polishing compounds come as either a hard waxy block or as a cream and are applied to the cast surface with lint-free cloth or a mechanically spun polishing mop.

Cloth polishing mops for block compounds and sponge mops for cream compounds can be mounted, using the correct arbour, to a standard electric drill with variable speed control, or mounted in a dedicated polishing machine. Polishing machines are bench-mounted, so will restrict the size of the cast that can be polished. Electric drills can be used freehand or securely bench-mounted using a clamping system specified to the machine to be used. Care should be taken when using electric drills to polish as they have no guards, and goggles should always be worn whichever system of polishing is being used.

Polishing compounds consist of very fine levels of abrasive and come in grades of coarseness. They should be used in sequence from coarse to fine methodically over the entire surface of the cast. Compounds usually come in two or three grades and castings should be washed between grades of polish. Cold-cast metal castings can be given a final polish using a proprietary metal polish suitable to the base metal used.

If polishing through the grades of compound fails to produce the levels of gloss, brightness or clarity required, it may be necessary to return to abrasive papers to cut back more material and then to repolish.

Tips

- When cutting back and polishing cold-cast metal castings with a lot of difficult to access detail, different grades of wire wool or very fine wire brushes can be used. These should be used dry, so care should be taken not to generate too much dust and a dedicated dust mask should be worn.
- Wire wool should not be used on white or light-coloured castings as they may discolour.
- When cutting back a casting with abrasives the object is to remove the surface layer of resin to expose colour or base materials. On castings with very fine detail this process may remove some detail. There is no way round this, so it should be considered how much cutting back is acceptable.
- Castings with areas of high and low detail will produce areas that are brighter on the high points. This can look very effective if a 'foundry sculpture' effect is required.
- Once cut back and polished, cold-cast metal castings will begin to oxidize over time but can be brought back to brightness with a proprietary metal polish.
- Cold-cast bronzes (and other metals) can be displayed as such, as long as they are labelled 'cold cast' to distinguish them from 'foundry cast'.
- Toothpaste is an ultra-fine abrasive, and can be a good final polishing compound!

Surface Treatments

- Cold-cast bronze, copper and brass castings can be 'paginated', following the same methods as for 'full' bronzes. Pagination times may need to be reduced due to the smaller amounts of metals contained within the resin (for pagination techniques, research other publications).
- Casts can be wax polished to retain brightness.
- Metal shellacs can be used for a permanent seal.
- Domestic metal polishes can be used to rejuvenate brightness on oxidized castings.
- Iron-filled castings can be submerged in water to produce a rusted bloom on the surface.

Faults and Repairs

- Small holes or cracks can be filled from the surface with additional mixes. GP resin and the appropriate metal powder are combined at a ratio of one part resin to at least two parts powder, producing a thick paste that is used to fill the repair. Overfill and cut back the excess.
- If possible, laminate behind repairs with fibreglass mat.
- Broken-off cast pieces can be glued back with GP resin containing a little metal powder or colour.
- Wax gilt crayons can be used to touch up minor blemishes.

CHAPTER 12

CIMENT FONDU CASTING

Although more familiar as a building material, concrete as a casting material can be a very versatile, accurate and above all durable medium. Early use of concrete at the beginning of the twentieth century to mass-produce garden ornaments introduced cheap castings that had durability to the elements, although the methods and results where crude. It was not until after World War II that mouldmaking and casting techniques were sufficiently developed to enable a wider adoption of concrete as a versatile and accurate casting technique.

Cheap and readily available concrete can be worked with a number of different surface treatments. Traditionally, plaster moulds are used but it is possible to cast in virtually any type of mould. It is usually preferable to create hollow concrete castings, in order to economize on weight and material. The method of casting concrete hollow is similar to the production of any hollow casting material – surface layers of cement are reinforced with a laminate, usually fibreglass matting, producing very strong, hollow castings.

Materials

Cement

There are many types of cement available and although theoretically any could be used to produce a casting from a mould, aluminous cement (ciment fondu) or Secar are preferable due to their fast setting time and good outdoor weather resistance. Fondu produces castings that are dark grey in colour and Secar off-white. Cement on its own has little tensile strength and will always need the addition of sand to produce full strength.

Sand

Different types of sand are available, producing different textures and colours.

Soft sand is quite fine, dark yellow in colour and is generally used with fondu as it produces darker castings. Soft sand usually contains a lot of water, sometimes too much for casting purposes and may need to be dried out before use. It is preferable to buy bagged soft sand rather than use sand from a pit that has been left to the elements and may contain excess and dirty water. A batch can be tested by putting a teaspoonful in a glass of water, which should only become slightly cloudy.

Silver sand is finer and much lighter in colour to soft sand and is preferable when using Secar to produce lighter castings, and is usually sold bagged and dry.

Sharp sand has a large particle size and subsequently will produce castings with a coarser surface texture.

Any combination of these three sands can be combined to produce different results.

Storage and Handling

Both sand and cement should be stored in airtight bags or containers in a warm dry place. Batches of cement that

OPPOSITE: **A fine example of a ciment fondu cast. Note the highly textured appearance.**

149

contain lumps should be discarded, as setting times and set strength may be compromised. If in doubt about any of the materials a test batch should be produced.

As with any fine mesh (powdered) material care should be taken to avoid clouds of airborne material that may produce a respiratory hazard.

Tools and Equipment

Following is a list of the tools and equipment needed for the various processes involved in the casting of cement.

- Builder's grade plaster bowls and buckets are the principal pieces of equipment needed for the measuring and mixing of ciment fondu. Because of the relatively coarse nature of the materials involved it is not advisable to use bowls and buckets that will be used for plaster work as they will inevitably become scratched, making them difficult to clean. For mixing large batches a shallow wooden box can easily be constructed from scrap timber.
- Measuring utensils, plastic cups or bowls (of a size appropriate to the batch sizes) to dispense volumes of cement, sand and water.
- Wooden paddles or batons are good for mixing small batches; a trowel or small shovel is useful for larger ones.
- Cement is a very hard material when set, so any trimming and cutting back of castings will need to be done with a mallet and chisel. Old wood chisels or stone carving chisels can be used to remove large amounts of excess flashing on seams.
- Rifflers followed by carborundum paper can be used for finer surface imperfections.
- Various sizes of decorating brushes are needed for the application of the goo coat.
- Sponges and a nylon brush are required for cleaning off excess cement from mould edges.
- Nylon scourers are useful for washing 'blooms' (a white deposit left by plaster moulds) from a cast.
- Tamping is the method of tapping fondu into the surface detail of a mould. A rounded tool handle can be used for this job, or a tool that is custom-made for the size of the job (made by rounding and sanding a piece of timber).
- A variety of scrapers and modelling tools are useful for scraping off the excess from mould surfaces.

- Scissors are required for cutting laminates.
- Hacksaw/pliers are needed for cutting armature material.
- A wire brush is needed for cleaning tools.
- A builder's bucket is required for washing equipment prior to sink washing.
- A hot-air gun, of the decorator's variety, is needed for softening wax locally.

Studio Practice, Personal Protective Equipment and Cleaning

- If excess cement is allowed to enter a drainage system it will set permanently, so a sink with a plaster trap is useful to avoid drain blockage (*see* plaster casting).
- Wash all tools and utensils before the cement is allowed to set.
- Tools and bowls should be rinsed in a bucket before sink washing, even in a sink with a plaster trap. A handful of casting plaster mixed into the washing bucket and allowed to settle before use will prevent the washed-off cement from setting on the bottom of the bucket. When washing has been completed allow the bucket to settle before pouring off the water and binning the waste cement.
- Kitchen paper towels or old rags are needed for cleaning tools and utensils.
- A dry, warm place is required for material storage.
- Work surfaces can be protected with heavy gauge polythene.
- If skin is sensitive, barrier creams should be applied and/or protective gloves worn.
- When mixing large quantities of cement keep the dust down to a minimum and wear a proprietary dust mask when necessary.

Measuring and Mixing

Standard Mix

The proportions of cement, sand and water in a standard mix should be kept consistent between batches, so that colour matching and integral strength are maintained. The proportions are the same for fondu or Secar. The standard proportions of a mix are as follows:

- cement   two parts
- sand   six parts
- water   one part.

1. Measure out the volumes with utensils that are appropriate to the size of the mix. Do not use the same measuring utensil for wet and dry material, particularly when making several batches of mix.
2. Ensure mixing bowls are clean. Old mixes may affect setting time and strength of subsequent batches.
3. Blend the dry ingredients thoroughly by hand or with a tool.
4. Add water bit by bit and mix into the dry ingredients.
5. Mix the batch thoroughly until an even colour and consistency are achieved.

N.B. Adjustments of water may be required; too much water in a mix will produce porous and weak castings, as once the water has evaporated air pockets will be left. The standard mix should be quite dry – it should be able to hold form without any water coming out if a handful is squeezed.

Goo Coat

The mix proportions described above are for a standard mix that can be applied straight to the mould, but to avoid air bubbles and ensure good detail capture it is beneficial to apply a 'goo coat' first, particularly on smaller sculptures with fine detail.

A goo coat is a mix of just cement with enough water added to obtain a consistency of double cream (the consistency should be brushable). This is then painted onto the mould surface prior to the standard mix being applied on top. Castings that are to be kept outdoors should have a goo coat of two parts fine sand to one part cement, and enough water to create a mix that is just brushable.

Colour

Different colours can be produced by using different cements and sands in varying combinations. As long as the proportions of cement, sand and water for the standard mix are adhered to, any combination of sands and cements can be used: for example, different combinations of fondu and Secar will produce varying shades of grey; Secar and silver sand will produce a pale cast; Secar and crushed marble will produce a white cast; and Secar and stone powders will have different stone effects. It is always a good idea to do a test batch to determine the desired colours.

Powder pigments can be added at the mixing stage to produce integral colour to a cast, and specially designed dyes and pigments can be purchased from a builder's merchant. Again, test-batching is a good idea.

Most important when colouring cements is to mix up enough dry material for the whole casting job to ensure colour matching between batches. When making up large mixes like this, always use dry sand and colouring material otherwise the whole mix will start to set before you can use it. Once the whole amount of dry material has been mixed, water can be added a batch at a time.

If the mould is primed with a thin red clay wash, this will becomes incorporated into the cast, giving it a red tinge.

N.B. When using mixes never allow them to become too dry. If the mix becomes too stiff it will not pick up detail or adhere properly to the mould. Mix up only enough material that can be used within an hour, and if in any doubt about the workability of a batch throw it away and make a new one. Never add more water to make a stiffening batch workable again, as this will compromise the set strength of the casting.

Principal Casting Techniques

Hollow Casts

Casting with ciment fondu is generally a four-layer laminating process consisting of, in sequence:

- the goo layer;
- the first standard mix layer;
- the fibreglass and goo layer;
- the second standard mix layer.

THE GOO LAYER

1. Apply the first goo layer with a brush, evenly and methodically to the mould surface to a depth of approximately 3–4mm.

2. Very large moulds will need to be completed in sections (no more than 0.5m²) so that layers do not dry too much before they can be laminated with the next layer.

3. This first layer is picking up the detail of the mould so it is important that it is thoroughly and evenly applied.

THE FIRST STANDARD MIX LAYER

1. Apply the standard mix layer immediately after the application of the goo layer.

2. Apply a tablespoonful of standard mix to an area of goo layer and spread out to a thickness of approximately 5mm initially. Spreading of the standard mix is best done with the fingertips and should create a wave of goo ahead of it as it spreads. This has the effect of displacing the goo as it spreads, which will fill all the mould detail and push out air bubbles as it goes.

3. It should be noted that the goo layer needs to be able to move while carrying out this stage, so as mentioned before an area of no more than approximately 0.5m² should be covered at a time. Continue to coat the goo layer and spread out the standard mix in this way until completely covered to 5mm thick.

4. In deeply undercut or detailed areas this thickness should be increased so that the mould interior is smooth. It should eliminate most of the undercutting and detail, making it easier to laminate with fibreglass mat later.

5. If working in a mould with an overhang, work up to a vertical point, allow to cure, then the next day rotate the mould and complete the uncovered parts.

N.B. The application of the standard mix layer to the Goo layer takes a little practice, and it is advisable to practise the technique before application to the mould (a non-porous flat surface is good for this).

Compacting and Curing The first Goo layer and standard mix layer applied to the mould, they need now to be compacted and left to cure. It is essential to compact these first two layers to eliminate air: this will produce a casting with the full potential strength of the concrete. Grains of sand are spherical which unless the cement is compacted to fill the spaces between them will leave air pockets, making the cast weak. This process is known as 'tamping', which is best described as a light hammering, and can be done with a 'round ended' tool handle or a custom made tool.

1. Lightly tamp the entire surface methodically, ensuring that it feels firm to the touch. This process is made much more efficient if the standard mix does not contain too much water (see mixing above).

2. Once tamped, allow these layers to cure for 24 hours before applying the laminate layers.

3. Curing (hardening) of cement needs to be carefully controlled to prevent too rapid a loss of water from the cast. The cast should be allowed to harden only until firm to the touch, but should not be dry, and at this stage the cast must be prevented from losing moisture too rapidly.

4. Spray the surface with water, and cover with damp rags to contain the moisture. The cast should be checked over the next 24 hours and rags sprayed repeatedly to maintain dampness. The important point with cement cure is not to allow the cement to dry too rapidly which will produce an irreversibly powdery and weak casting. The cast is now ready for lamination with fibreglass.

Supporting the Mould While Curing If a plaster mould is being used it will need to be efficiently supported while the cure takes place. Because the cast needs to be kept damp for up to 24 hours while the cast cures, plaster moulds may warp, particularly piece moulds that may have to be laid up in separate pieces.

This problem can be greatly avoided at the mouldmaking stage by reinforcing moulds with armatures. One-piece moulds with a wide opening can have temporary tie bars attached across the opening. Moulds in pieces will need to fully supported under any overhangs with blocks of wood or bricks.

Flexible moulds with a case, being non-porous, will not need to be supported as the moisture will not be allowed to penetrate through to the case.

THE FIBREGLASS AND GOO LAYER

The material used to reinforce castings is 300g fibreglass mat; this is also the type used in polyester resin casting and is made up of strands of fibreglass formed into a mat. Fibreglass mat is fairly rigid, but will soften a little when dipped into a goo mix, although not enough to get into deep undercuts or very fine detail. It is therefore important that the previous layers are applied thickly enough to eliminate as much undercutting and detail as possible.

1. Cut up enough relevant-sized pieces of mat to cover the cast (pieces that are too large will bunch up; too small and they

will take an unnecessary amount of time to apply).

2. A layer of goo should be applied to the cast, which should be hard but still damp from the wet rags. The mat is then dipped in goo mix and applied. The mat should be thoroughly saturated with goo and pressed onto the surface to eliminate air bubbles and ensure good lamination.

3. Continue to cover the entire surface, overlapping mat pieces by a few centimetres.

Additional Reinforcement

■ Large casts may need to be reinforced more substantially using additional layers of fibreglass, steel rods or timber batons.

■ If the shape and depth of the cast permits (such as down the legs of a standing figure) steel rods can be embedded into the mix, which should be done at the first standard mix stage and should be at least 12mm thick.

■ Other shapes may benefit from surface-mounted reinforcement (for example, across the concave surface of a torso), which should be done during the fibreglass layer.

■ Ensure that a good layer of fibreglass and cement holds each end of the rod or baton in place securely.

■ Steel rods should be painted with red oxide before use to prevent them rusting.

THE SECOND STANDARD MIX LAYER

The goo mix incorporated in the fibreglass laminate layer contains excess water that should be adjusted now by application of a final standard mix layer with half the usual amount of water. This will produce a mix that can be sprinkled onto the surface to suck up excess water. This should then be tamped and wrapped in wet rags as for previous layers. This produces a sandwich to the laminate layer.

The cast is now complete and should be allowed to cure, ensuring that it is kept damp and the mould supported as before for 24–48 hours until fully hardened.

Solid Casts

It is possible to fill small or relief moulds solid, but be aware of the weight and amount of material needed.

1. Initially treat the mould with a goo layer as for hollow casting.

2. Make up a standard mix with a little extra water, so that it is slightly sloppy, and add this a little at a time, while tapping the mould vigorously to expel air bubbles.

3. As the mould is tapped, air bubbles will be seen to rise to the surface along with some excess water, which should be sponged off the surface before more mix is added.

4. Continue this process until the mould is filled and allow it to cure fully with damp rags as for other castings.

Laying-Up Piece Moulds

Moulds that have been made in pieces will need to be filled with cement while still in pieces, then joined.

■ Initially, the mould should be assessed to determine how many pieces can be assembled while still allowing access to fill it. It may be that some moulds can be assembled excluding one piece, which can be filled separately and then put in place last. Other moulds may need to be in separate pieces and then assembled for joining.

■ When making a cast in pieces particular attention needs to be paid to the edges that when joined will become the seams of the cast.

■ Where seams can be reached by hand to join them, the cement layers should be built up to the edge of the mould pieces, left slightly rough and then slightly chamfered back (angled down from the edge of the mould). The seam can then be built up with cement and laminated when the pieces have been assembled.

■ When joining the cast pieces the same layering procedure should be followed as for the rest of the cast and should extend 25–30mm either side of the seam.

■ Where seams cannot be reached by hand when the mould is assembled, the cement layers should be built up just below the edges of the mould pieces and left slightly rough.

■ A thick mix of goo is now made (five parts cement to one part water) which is placed with a palette knife around the seam edges to be joined. The mix should sit as a bead proud of the seam edges. The mould pieces are then squeezed together, pushing the beads of goo together. Some goo will be squeezed inside the cavity of the hollow cast, and some will be squeezed outside of the mould piece seam, which should be wiped off before it dries. The mould pieces should be squeezed and tapped together thoroughly and methodically around the entire seam to ensure a good bond and the mould pieces securely fixed together.

- It should be noted that whichever method of joining a mould in pieces is employed, when filling the pieces with cement layers it is very important not to allow cement on to the mould seams. This will present inaccurate registration when assembling the mould for joining. Any excess on the mould edges should therefore be sponged off as you go.

Demoulding

The techniques of demoulding will largely depend on what type of mould is being demoulded.

- **Plaster one-piece** Moulds can be tapped or levered out from the *inside* of the cast using a chisel. Soaking in hot soapy water can help to ease out difficult casts.
- **Plaster multi-piece** Moulds should be eased apart piece by piece. Small moulds can usually be parted using a strong-bladed knife. Larger moulds may require a thin chisel to crack the seam initially until a wooden or plastic wedge can be driven in. On particularly long seams, a series of wedges can be gradually driven in until the seams part.
- **Plaster waste** *See* chipping out plaster waste moulds.
- **Fibreglass** As above, although there can be a certain degree of flexibility in a fibreglass mould depending on how thickly it has been made.
- **Flexible moulds (without a case)** Demoulding from flexible moulds without a case is a combination of flexing and peeling the mould while easing the cast out.
- **Flexible moulds (with case)** First remove the case (this should be carried out as for a plaster piece mould) and then peel the mould away from the cast. It is possible to peel the mould away from the cast with these moulds as they are thinner and more flexible than flexible moulds without a case.

CASTING A TWO-PIECE RUBBER MOULD

Application of the goo coat.

The second standard mix layer is applied by hand.

The mix layer is tamped with a round-handled tool.

The goo and mix layers applied, the cast is wrapped in damp bandages to set for 24 hours.

Applying the fibreglass laminate back-up layer. This is backed up with another mix layer and tamped as before.

A bead of goo mix is applied to the seam before joining the mould pieces.

CASTING A TWO-PIECE RUBBER MOULD *continued*

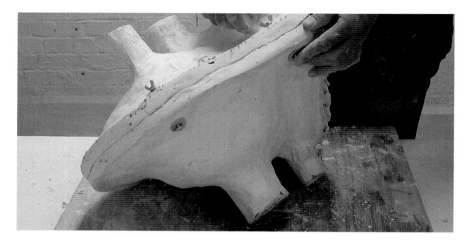

The mould pieces are joined and bolted together. Damp bandages are applied and the cast is allowed to set fully before the mould is opened.

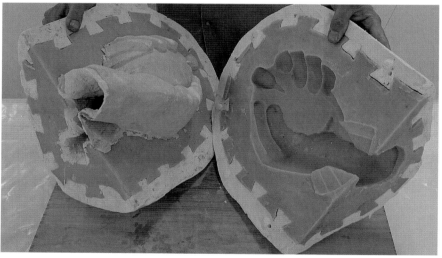

The mould pieces are parted.

The vinyl is peeled away from the cast.

The finished cast.

Trimming and Cleaning

- Cement is an extremely durable material and will need a mallet and chisel to trim off any flashing (excess material that has leaked between the seam of a mould). Stone chisels or old wood chisels for smaller seams are good for this job.
- When closer to the true surface of a cast, steel rifflers can be used, and finally finer and finer grades of carborundum paper.
- When casting ciment fondu in plaster moulds the only release agent that is required is a thorough soaking in water. However, the cement tends to deteriorate on the surface of the mould, producing a white residue on the cast surface known as a 'bloom'. This can either be left (it can look quite attractive) or removed from the casting using water and a nylon brush or scourer.

- When chipping out casts from plaster waste moulds, it may be desirable to stain any remaining pieces of plaster that cannot be removed.
- Flexible moulds may leave a greasy residue that can be removed with white spirit followed by hot soapy water.

Surface Treatments

Although the surface of fondu and Secar castings can be very attractive if simply scrubbed down with detergent and left untreated, a number of different surface treatments can be applied.

- One of the simplest and most effective is the application of a wax polish. Standard furniture polishes can be used following the manufacturer's instructions.
- Acrylic or polyurethane varnishes can be applied.
- A bronze-like effect can be achieved by vigorously brushing the cast surface with a fine brass-wire brush. The brush will leave a deposit on the high points of the casting.
- Metal powders suspended in polyester resin can be applied and cut back with abrasives to reveal the metal.
- Any type of paint can be applied and then waxed or varnished. Cast surfaces may need to be primed prior to some types of painting (check the manufacturer's instructions).

N.B. It should be noted that colour testing should always be carried out on an obscure part of the cast before commitment. Application of paint and resin treatments may result in the loss of very fine detail from a cast surface. Weather resistance of surface treatments should be considered on castings that are going outside. Weather resistance is generally good on castings with no surface treatment as long as mix proportions are adhered to and kept consistent.

Faults and Repairs

- Small holes and cracks can be filled with small batches of ciment fondu. The area to be filled should be thoroughly soaked with water before filling.
- Large breaks can be glued with epoxy resin glue. Cast pieces should be thoroughly dry before gluing.

CHAPTER 13

WAX CASTING

Wax has been used as a modelling medium for thousands of years, pre-dating the use of bronze. Many Renaissance artists used wax to produce maquettes as preliminary studies prior to producing larger work in other media. Development of mouldmaking and casting processes over the centuries has enabled artists to produce wax casts of everything from life casts ('waxworks'), to anatomical models and artificial fruit. One of the most important uses of wax as a casting medium is in the production of bronzes, known as cire perdue, or the lost wax process. The process involves the use of a modelled or cast wax, around which a refractory mould is made, then the wax is melted out and bronze poured in its place. The method can be dated back at least 4,000 years and is still used today (*see* Chapter 1).

There is a wide variety of modern waxes available for casting purposes either to use as a medium in its own right or as the first stage towards the production of a bronze. It may be of interest to mention that the production of a mould, to then produce a wax, is at least half the job in the production of a metal casting, and presenting a foundry with a finished wax will considerably lower the production cost.

Wax can be coloured, easily trimmed or remodelled after casting and can be used in a variety of different moulds.

Materials

There is a wide variety of waxes available, some of which are more suitable than others for casting in moulds. Different combinations can produce waxes of different colours, hardness and workability. When combining waxes, test batches are essential to determine which combination is most suitable for your purposes.

Microcrystalline Wax

Usually supplied in blocks or slabs. It is non-toxic, odourless and very fluid when melted, ideal for pouring and painting into moulds. This is a multi-use wax that is flexible, non-staining and will combine well with other waxes.

Melting point 71°C.

Paraffin Wax

Usually supplied in pellet form and refined to eliminate impurities. It is primarily used for candle making and modifying other waxes, and is relatively soft.

Melting point 57°C–60°C.

Green Casting Wax

Primarily designed for the lost wax metal casting process. Quite firm at room temperature, it can be carved, modelled when warmed in the hands or mixed with other waxes. It can also be poured or painted into moulds with good results.

Melting point 63°C–66°C.

Modelling Waxes

As the name suggests, these are blended waxes designed primarily for modelling, but they can be used for casting purposes. Generally available in different colours that denote their hardness, they can be mixed together or with other casting waxes to produce intermediate hardnesses or colours.

Melting point 52°C–74°C.

Machining/Carving Wax

A very hard wax for machining and carving. It can be cast but has a 6.5 per cent shrinkage.

Melting point 120°C.

Beeswax

A very soft wax used primarily as an additive to soften other waxes.

Melting point 61°C–65°C.

Glass Wax

A clear wax that is very brittle. Used in the film industry to produce fake glass. It is best poured into silicone moulds.

Melting point 115°C. Pouring temp 120°C–140°C.

Rosin

An additive that is melted into waxes to add hardness.

Tools and Equipment

Following is a list of the tools and equipment needed for the various processes involved in the casting of wax.

- Melting equipment is the principal piece of equipment needed when casting in wax, and a heavy-based saucepan with a lip can be used on a domestic cooker very effectively. A cooking thermometer is useful to determine exact temperatures. Alternatively, some form of thermostatically controlled device makes the job a lot easier. Commercially available electric frying pans with a controllable thermostat built in are very good either to melt directly in, or to heat a separate pan. If the pan does not have a lip, it is a fairly straightforward job to knock one out with a small hammer, and is essential if pouring directly from the pan.
- Brushes (decorating quality) are needed for the application of wax to the mould surface when hollow casting.

- Wax modelling tools are required for remodelling and trimming casts.
- Craft knifes are needed for trimming flashing and seams, and are also useful for welding repairs.
- An alcohol burner is needed for heating tools when trimming and welding (a candle or any safe naked flame can also be used).
- A builder's bucket is needed for rapid cooling of small moulds and pouring off excess wax for reclamation.
- Kitchen paper towels or old rags are required for cleaning tools and utensils.

Studio Practice, Personal Protective Equipment and Cleaning

- Good ventilation is vital. Wax will give off minimal fumes unless overheated, but it is still necessary to melt wax in a well-ventilated area to prevent a build-up of fumes.
- Wax has a low flashpoint (the temperature at which it will combust), so it is important to have some fire-fighting procedures in place – an extinguisher (foam or relevant for wax fires, not water) and a fire blanket. Never try to extinguish a wax fire with water, as it may explode.
- Long-sleeved overalls and rigger's gloves (heat-protection specific) should be worn for protection against splashes.
- Heat-protection specific goggles should be worn.
- Workbenches should be protected with several layers of damp newspaper, but ensure that there is no direct contact with any electrical equipment being used. Melamine benches can be scraped (but do not scrape decorative melamine work surfaces!).
- Excess wax in a saucepan can be poured off into a bucket or sink of water to be reclaimed, but allow it to cool until skinned over before pouring.
- Wax should be stored in a cool dry place and wrapped or bagged in strong polythene to prevent dirt and dust contamination.
- Tools can be heated just until the wax starts to melt and then wiped clean with kitchen paper towels or old rags.
- Pans or melting equipment can be cleaned by pouring off the excess, allowing it to cool slightly and then wiping the pans out with a paper towel or cloth.

Melting

Wax in large blocks should first be cut into square cubes of approximately 50mm. This can be done with a saw or very carefully with a knife. Pans or other melting equipment should be clean and absolutely dry.

Initially melt just a couple of cubes of wax until it is liquid. Add more blocks gradually until the required amount has been melted. Do not overfill melting utensils, as they become dangerous to handle when pouring. When melting wax it is important not to overheat it, so whether melting on a stove top or in a thermostatically controlled pan it is important to regulate and check the temperature as the wax is melting. Obviously with a thermostatically controlled pan the temperature can be set to the melting point of the wax being used, but on a stove top a pan must be checked regularly with a cooking thermometer. Start at the lowest setting and increase gradually until the wax just begins to melt. Continue at this setting, checking the temperature regularly to ensure it is not getting too hot. A good guide is that if the wax starts to smoke the melting temperature is too high. If melting a pan with solid wax left over from a previous job, place just one side over the heat until a channel has melted through to the surface. Then place the whole pan over the heat. This will avoid a bubble building up under the solid surface wax.

Wax is a petrochemical with a low melting temperature and flashpoint (temperature of combustion), so great care should be observed in melting procedures (never put out a wax fire with water. If possible, remove the pan from the heat and smother).

Once the required amount of wax has been melted the temperature should be regulated to either the pouring, painting or slushing temperature (see below).

Colour

Wax can only be coloured integrally due to its repellent nature. Colour has to be added to wax in its melted state, and wax pigments are available specifically for this purpose. The pigment is suspended in a carrier wax that enables it to be readily mixed into the melted casting wax. Mix in a little pigment at a time until the required colour is achieved. Alternatives for colouring wax are wax crayons, which can be melted into a mix in chunks, or canvas oil paints, which can be added in squirts until the colour is achieved.

It is a good idea to test coloured waxes, allowing a small amount to cool out of the pan to check the colour before committing to the casting.

Pouring

Solid Casts

To make a solid wax cast it is possible simply to pour the melted wax into a mould, but some factors need to be considered before doing this. For example, the type of mould and whether or not release agents need to be applied should be considered (see Chapter 3). A solid casting will obviously use more material so this may also be a consideration, in terms of cost or melting out of a lost wax mould, for instance. More material in a mould will take longer to cool down and set. There is a certain amount of shrinkage as wax cools, which, unless the pouring hole or mouth of the mould is extended, will form an indent at the point on the cast where the mould has been filled. The pouring hole or mouth of a mould can be extended with either a funnel or a clay wall to make it higher, which will allow some space for the wax to shrink without shrinking into the cast. At this point there will be a little excess wax that can be trimmed when the cast is demoulded.

1. Ensure that the mould is stable and level on the work surface, and that the pouring hole or mouth of the mould is at a level that is easy to reach with the pan of wax. Large moulds may need to be set up on the floor.
2. Melt the required amount of wax to its melting temperature and then allow it to cool until the surface of the wax just loses its shine or leaves a slight line on the edge of the pan when tilted. The pouring temperature needs to be cooler so that it sets instantly against the mould surface. Carefully fill the mould in one steady pour to the highest level of the pouring aperture.
3. Allow the cast to set fully before demoulding. This will depend on the size of the mould and the wax being used and can be speeded up by immersing the mould in cold water after the wax has set firm enough to move the mould safely.

Hollow Casts (Slushing)

Hollow castings have the advantage of using less material, making them more economical and lighter in weight. The method of pouring wax to create a hollow cast is known as 'slushing', and involves building up layers of wax by filling the mould, pouring out the excess and cooling each layer until the thickness required has been built up. This method is suitable only for moulds up to a certain size, as they need to be lifted and poured out safely several times.

1. Procedures for preparation and filling the mould should be followed as for the solid cast method above. It may not be necessary to build up the pouring aperture quite so high as the shrinkage will be less.
2. The mould does not need to be filled fully – about half full will be adequate, at which point it should be carefully tilted and rotated, slushing the contents to cover evenly the whole mould surface.
3. Pour the excess wax back into the heating utensil, and allow the first wax coating to set and cool.
4. This first layer will create a deposit of wax 1–2mm thick. Subsequent layers should now be built in the same way, allowing them to cool between layers, until a total thickness of approximately 6mm has been achieved.
5. Piece moulds can be filled with this method; all the mould pieces should be securely joined to ensure there is no leakage when tipping out excess. The seam lines between pieces can be reduced to a minimum using this method.

Brushing

Hollow wax casts can also be created by building layers by brush. This method is necessary when casting from piece moulds that need to be filled open in separate pieces and then joined. This method needs to be followed if the mould is very detailed and undercut, potentially creating problems with air bubbles if the wax is poured, or if the mould is very large. If filling a one-piece mould there will obviously need to be enough access at the mould opening to be able brush the wax onto the whole surface of the mould.

1. Preparation of the mould and melting of the wax should be followed as above. If filling a piece mould, join as many pieces as possible while still allowing access to the mould surfaces. The temperature of the wax pan needs to be lowered to the same degree as for pouring and maintained at that temperature throughout the job.
2. Carefully paint a first layer on the mould surface, reloading the brush from the wax pan after each brushful. If the brush starts to harden while painting leave it in the wax pan until soft. Stippling the wax using the end of the brush to apply the wax can produce a more even surface finish. It may be necessary to experiment depending on the size of the mould and the required surface finish.
3. If filling a mould in separate pieces, make sure to keep within the edges of the piece. If wax gets on to these edges it will make it difficult to join the mould. Chamfering (angling down from the edge of the mould) the wax away from piece edges will also aid reassembly of the mould.
4. The first touch of each brushful will deposit a 1–2mm layer of wax on the mould surface. This is enough for the first layer, which should now be allowed to set and cool thoroughly.
5. Build up subsequent layers in this way until a thickness of approximately 6mm has been achieved.
6. If filling very large moulds, more layers can be added or even a layer of hessian scrim, which should be thoroughly soaked in wax and sandwiched within the last two layers. When laminating with a scrim layer, push out air bubbles with a spatula and ensure they do not overlap piece mould edges (do not use scrim on lost waxes, as the casts would not burn out cleanly).
7. If filling piece moulds, the pieces should preferably be joined as soon as possible to ensure good adhesion between them.

Joining Cast Pieces in Piece Moulds

Mould pieces that have been filled separately now need to be reassembled in order to join the cast pieces within. This can be done in two ways depending on the amount of access there is into the mould once the mould pieces are assembled. If there is access by hand, wax can be painted over each seam and built up in layers and reinforced with scrim if necessary. If there is no access by hand, wax must be poured into the pouring hole and slushed several times to join the seams.

ACCESS BY HAND

1. Ensure that all the mould edges are clean of wax and that the cast edges are slightly chamfered away from the seam.
2. Securely assemble the mould pieces.
3. Apply a layer of wax to all the seams by brush.
4. Allow the wax to set and cool.
5. Repeat until all the seams have been covered to a little thicker than the thickness of the cast (reinforce with scrim if necessary).

NO ACCESS BY HAND

1. Ensure that all the mould edges are clean of wax and that the cast edges are slightly chamfered away from the seam.
2. Securely assemble the mould pieces.

3. Half fill the mould with wax.

4. Slush, ensuring maximum coverage of the seams.

5. Pour out the excess wax.

6. Allow the wax to set and cool.

7. Repeat three or four times to build up the wax thickness over the seams.

Combination Technique

This technique combines painting and pouring wax to join the cast seams.

1. Ensure that all the mould edges are clean of wax and that the cast edges are slightly chamfered away from the seam.

2. Join as many mould pieces as possible while still allowing access by hand.

3. Apply a layer of wax to all the seams by brush.

4. Allow the wax to set and cool.

5. Repeat until all the seams have been covered to a little thicker than the thickness of the cast (reinforce with scrim if necessary).

6. Join all the remaining mould pieces, ensuring that the edges are clean and chamfered.

7. Half fill the mould with wax.

8. Slush, ensuring the maximum coverage of seams.

9. Pour out the excess wax.

10. Allow the wax to set and cool.

11. Repeat three or four times to build up the wax thickness over the seams.

12. With all joining methods, allow the casts to set fully and cool before demoulding.

13. It is also possible to join casts made in separate pieces and then removed from a mould using 'wax welding' (see Faults and Repairs).

Demoulding

The techniques for demoulding, that is, removing the finished cast from the mould, will largely depend on what type of mould is being used.

- **Plaster one-piece** Moulds can be tapped and casts levered out gently from *inside* the cast with a blunt clay

tool. Be careful not to puncture through hollow casts with the tool. Sometimes casts can be floated out in a sink or bucket of lukewarm water as long as they are not too undercut.

- **Plaster multi-piece** Moulds should be eased apart piece by piece. Small moulds can usually be parted using a strong-bladed knife. Larger moulds may require a thin chisel to crack the seam initially until a wooden or plastic wedge can be driven in. On particularly long seams, a series of wedges can be gradually driven in until the seams part.

- **Plaster waste** *See* chipping out of plaster waste moulds, p.35. Be extremely careful when chipping out waxes from waste moulds as the surface can be easily damaged. It will be necessary to build up a very thick cast or use a very hard wax when using a waste mould.

- **Fibreglass moulds** As for plaster one-/multi-piece moulds, although there is a certain degree of flexibility in a fibreglass mould depending on how thickly it has been made.

- **Flexible moulds (without a case)** Demoulding from flexible moulds without a case is a combination of flexing and peeling the mould while easing the cast out.

- **Flexible moulds (with a case)** First remove the case (this should be carried out as for plaster one-/multi-piece moulds) and then peel the mould away from the cast. It is possible to peel the mould away from the cast with these moulds as they are thinner and more flexible than flexible moulds without a case.

- **Press moulds (without a case)** Essentially as for flexible moulds without a case, except that the clay can be wasted, that is, peeled or picked away in pieces, destroying the mould.

- **Press moulds (with case)** As for flexible moulds with a case, the clay usually being wasted in the process of demoulding.

- **Life moulds (mod roc)** Moulds should be thoroughly soaked in lukewarm water and then picked and peeled away from the cast.

- **Life moulds (alginate with mod roc case)** The case should be removed as for plaster one-/multi-piece moulds and then the alginate peeled and picked away from the cast.

HOLLOW CAST FROM A PLASTER PIECE MOULD

Join as many mould pieces as possible
while still allowing access.

Paint on the first layer of wax.
This layer is built up to thickness in
several layers.

Clean the seam edges of wax before
joining the last mould piece.

Join the last mould piece.

Wax is poured into the joined mould to seal the cast seam.

The mould is rotated to cover seams and the excess wax poured out. This is repeated several times to build up the seam.

HOLLOW CAST FROM A PLASTER PIECE MOULD *continued*

The mould pieces are parted, the cast removed and the seams trimmed.

The finished cast.

Trimming and Cutting Back

- Initially, any flashing on the seam lines or any excess can be trimmed away as close as possible to the cast surface with a craft knife or wax modelling tool. Trimmed excess can be saved and re-melted for other castings.
- When trimming right to the surface of the cast the tools can be heated slightly over an alcohol burner to achieve a really smooth surface.
- Finally, some gentle rubbing with the fingers over the seam lines should completely remove any imperfections (a light blast with a hairdryer can aid this process).

Surface Treatments

There are no surface treatments that can be applied to a wax cast surface that will protect it, due to the nature of the material itself being a repellent. However, if the surface of a cast becomes dirty or dusty it can be wiped down gently with white spirit.

Faults and Repairs

Inevitably some castings may not come out of the mould perfect. There may be cracks where the cast has shrunk around an undercut or there may be seams that have not joined properly.

- Where seams have not come together the wax can be 'welded' together. Welding involves melting a little of the wax on either side of the crack to be joined, which pools together and resets to form a strong integral join. This is achieved by heating both sides of a blade and introducing it briefly between the two surfaces to be joined. The blade is then withdrawn while applying equal pressure to both bodies of wax to produce the bond. Welding should not be confused with using the wax as a 'glue', which will produce a bond but not a very strong one. It is important to melt wax on either side of a join which can then run together to form a strong integral weld. Large wax-welding jobs may require the addition of 'wax rods' as the seam is being welded. A thin sheet of wax (3–4mm thick at the most) should be poured onto a thoroughly soaked plaster slab, allowed to set and then cut into 'rods' 3–4mm wide. Rods can be fed onto a heated blade while welding to add extra material to the join. By using wax welding it is possible to join whole casts that have been cast in several pieces and then removed from the mould . A small soldering iron can be used as a safe and effective welding tool.
- Small holes or hairline cracks can be filled with a little hand-softened wax, and then rubbed back with the fingertips. Small, raised pimples caused by air bubbles in the mould can be rubbed off the cast surface with the fingertips.

**Cast wax figures. Naoko Kitamura.
Approximately 30cm.**

'Sphere'. By Nick Brooks. Plaster cast with graphite and iron powder finish. 25cm.

CHAPTER 14

ALTERNATIVE AND ADDITIONAL MOULDMAKING AND CASTING IDEAS

This concluding chapter covers briefly some additional and alternative mouldmaking and casting ideas that are too numerous to include in detail in this publication.

Mouldmaking and casting is an artform with multiple possibilities. Many of the methods and techniques are interchangeable and the possibilities for generating alternative ideas are as variable and personal as the individual creating them. Combining different techniques and materials and reworking materials from the original patterns can produce a fresh creative process in its own right. Sometimes, just being aware of the possibilities of other materials and techniques can generate creative ideas not considered before. There are also alternative materials and techniques that may be beneficial in terms of time, cost, safety or the facilities available.

Although there is not enough scope in this book to cover the multiple possibilities of alternative methods in depth, it may be of interest to list a few, for the reader to research further.

Mouldmaking

- **Reworking mould surfaces** It is possible to rework a mould surface once completed, thus changing subsequent castings from the original pattern. For instance, a plaster mould surface could be scratched or chiselled into, or a flexible mould can be cut into with a scalpel, producing raised lines on the surface of a casting.

- **Additions to mould surfaces** Additions to a mould surface will change subsequent castings from the original pattern. Clay additions on the mould surface will change the form of castings from the original pattern, although not permanently – the clay would need to be set up for each subsequent cast. This can produce an interesting series of castings, all slightly different but with an overall underlying similarity in form.

- **Latex** Latex rubber is a flexible moulding material suitable for 'skin' moulds. It is a liquid rubber that is painted on a pattern and built up slowly in layers to form a skin that can be peeled away from castings.

- **Life moulding silicone** This is a relatively new material that can produce long-life moulds from the body from which multiple castings can be made.

- **Jesmonite** A water-based acrylic alternative to polyester resins and fibreglass. The advantages are that it is non-toxic, has no hazardous fumes and the tools can be washed in water.

- **Gelatine** This was the first flexible moulding system. It is water-based and suitable for casting plaster casts.

- **Silicone sealant** Sealant, of the domestic DIY variety, can be extruded over a pattern, allowed to set and then cast into. It is only suitable for small moulds. Test for releasing capacity and detail pick-up.

- **Hot-glue-gun glue** Can be used in the same way as silicone sealant. Test for the heat capacity of the pattern material.

■ **Plasticine** Can be used as an alternative to clay when press moulding.

■ **Found ready-made moulds** Almost any hollow vessel could theoretically be used as a mould as long as the casting material can be removed from it. Check for undercutting and release agent requirement. Some examples are: moulded plastic packing material with cavities, Tupperware containers, plastic egg boxes, moulded vinyl floor tiles, indents made into a bag of clay, jelly/cake moulds, etc.

■ **Wood moulds** Geometric moulds can be made out of sheet wood. MDF will produce very smooth mould surfaces; grained plys and particle boards will produce surface detail on castings. The appropriate release agents will be required, depending on the casting material. Small moulds can be tied together with rubber bands, string or wire. Larger moulds can be screwed together for secure assembly.

■ **Cardboard/foam board** Can be used in a similar way to sheet wood moulds. It is only suitable for small moulds.

■ **Paper mâché** This can be purchased as a fine powder pre-impregnated with glue, which, if allowed to set over a pattern, will form a lightweight mould for small castings.

■ **Catering grade silicone** Used for mouldmaking that needs to be food safe, such as when casting chocolate, jelly, etc.

■ **Putty silicone** A two-part system that produces a putty-like mass that can be useful to take moulds from architectural detail and in situ immovable patterns.

Casting

■ **Reworking castings** Any casting can be reworked after it has been removed from the mould. Plaster and cement castings can be carved with chisels, resins can be ground or sanded to change their form (personal protective equipment and ventilation is an important consideration, and vital if reworking polyester resin as its dust is carcinogenic). Wax castings can be manipulated with a hot knife blade.

■ **Surface treatments** Most castings can be painted provided that appropriate primers are used. Metal powders mixed with PVA glue can be applied, cut back

with wire wool and polished to produce metal finishes. However, any surface application will compromise some detail on a casting and this should be considered before application.

■ **Internal armatures** Very large castings will require reinforcment from within. Timber, square section aluminium wire, steel rod or section can be used to span across large sections of a casting and should be securely built into the casting during construction. Chicken wire, steel or aluminium mesh can be laminated into castings to provide great strength. Ferrous metals should be painted with red oxide before application to water-based castings to prevent rusting.

■ **Combining several castings** Combining several castings as one can make an alternative creative statement in its own right. Separate castings can be joined with more of the same casting material being used or glued together with adhesives once fully dry.

■ **Food** This can be cast from food-safe silicone moulds.

■ **Silicone and vinyl** Although commonly used as moulding materials, these can also be used as casting materials in their own right. They can be coloured with polyester resin pigments.

■ **Low-melt alloys** These are white metal alloys of tin and lead that can be melted at low heats and cast in specific silicone moulds. Usually only suitable for small-scale castings such as toy soldiers and cars.

■ **Ice** Ice casts are possible in any mould that is watertight and if you have access to a large enough freezer. Rubber moulds are probably most suitable as they can be sealed most efficiently; plaster moulds should be sealed with waterproof varnish and a bead of petroleum jelly placed around the mould flange before assembly to prevent leakage.

■ **Soap** Recipes for making soap are available from libraries and the Internet and can be cast in most moulds.

■ **Wax casts for foundry casting** Wax casts can be produced with a view to foundry metal casting. Foundries presented with an original pattern in wax can use them to produce metal castings using the 'lost wax' process. Producing your own original pattern in wax can dramatically cut the cost of foundry casting.

■ **Papier Mâché** This can be purchased as a ready-mixed powder that can be compacted into a mould to pick up detail.

■ **Jesmonite** An acrylic water-based alternative to

polyester resins. Cold-cast metals, stone and wood finishes can be achieved. It is non-toxic, there are no fumes and tools can be washed in water.

▓ **Polyurethane resins** A non-styrene-based system of resins used in the same way as polyester resins but without toxic fume emissions.

▓ **Clay** Can be pressed into a rigid mould and then fired.

▓ **Clay slip** Liquid clay is poured into plaster moulds to produce ceramic castings that can then be fired in a kiln. Plaster moulds should be made thickly to absorb the water from the slip, leaving a deposit of clay on the mould surface.

▓ **Sawdust** Mixed with PVA glue, sawdust can produce wood-like castings.

▓ **Hanging fixings** Hooks, loops, wire or any method of hanging fixings should be securely built into castings that are to be wall mounted during construction of the casting. Stainless steel or galvanized fixings should be used on plaster castings to prevent rusting.

▓ **Resin-bound aggregates** Almost any material can be used as an aggregate, which can be saturated in clear casting polyester resin and packed into a mould, giving the impression of a casting made out of the aggregate. Some examples could be: glass chips, pebbles, gravel, ball bearings, marbles, beads, wood chips and so on. Testing for an adverse reaction to the resin is a good idea. Saturate the aggregate in the resin, pack it into mould and then top it up with more resin to bond the aggregate completely together.

▓ **Chavant clay (non-sulphurated plasteline)** This is an oil-based sculpting clay that can be mettled and poured into a mould. Non-drying (like plasticine) castings could be reworked and re-moulded if required.

▓ **Expanding foam** Available as an aerosol or a two-part liquid compound that sets to a lightweight firm cellular foam. It has minimal detail pick-up, so is really only suitable for large castings. Can also be used as a lightweight filler for hollow castings in other materials. Do not overfill; multi-piece moulds should be securely fixed together.

'Spheres'. By Nick Brookes.
Left: pressed clay/sawdust fired.
Right: plaster cast with graphite powder finish. 25cm.

MANUFACTURERS AND SUPPLIERS

4D MODEL SHOP
Arches, 120 Leman St
London E1 8EU
Tel: 0207 264 1288
Website: www.modelshop.co.uk

ALEC TIRANTI LTD
70 High St
Theale
Reading
Berks
RG7 5AR
Tel: 0118 930 2775
Fax: 0118 932 3487
27 Warren St
London W1
Tel: 0207 636 8565
Fax: 0118 932 3487
Website: www.tiranti.co.uk
(General mouldmaking and casting/sculpture supplies)

ACC SILICONES LTD (formerly Ambersil)
Wyldes Rd
Castlefield Industrial Estate
Bridgewater
Somerset
TA6 4DD
Tel: 01278 411411
Fax: 01278 727 266
Website: www.acc.silicone.com
(Silicone and vinyl rubbers)

BENTLEY CHEMICALS LTD
Rowland Way
Hoo Farm Industrial Estate
Kidderminster
Worcs
DY11 7RA
Tel: 01562 515 121
Fax: 01562 515 847
Website: www.bentleychemicals.co.uk
(Silicone rubbers/polyester resins and sundries)

BIG HEAD BONDING FASTENERS LTD
Unit 15
Elliott Rd
West Howe Industrial Estate
Bournemouth
Dorset
BH11 8LZ
Tel: 01202 574 601
Fax: 01202 578 300
(Embedding, joining and hanging fixings)

BRITISH GYPSUM
Newark Works
Bowbridge Lane
Newark
Nottinghamshire
NG24 3BZ
Tel: 01636 670249
Fax: 01636 670229
(Casting plasters)

EURORESINS UK LTD (HEAD OFFICE)
Cloister Way
Bridges Rd
Ellesmere Port
Cheshire
CH65 4EL
Tel: 0151 356 3111
Fax: 0151 355 3772
Website: www.euroresins.com
(Polyester resins and sundries)

FIBREGLASS SHOP
197 High St
Brentford
Middx
TW8 8AH
Tel: 0208 568 7191
(General mouldmaking and casting supplies)

JACOBSON CHEMICALS LTD
Jacobson House
The Crossways
Churt
Farnham
Surrey
GU10 2JD
Tel: 01428 713 637
Fax: 01428 712 835
Website: www.jacobsonchemicalsltd.co.uk
(General mouldmaking and casting supplies)

POTH HILLE & CO
37 High St
London E15
Tel: 0208 534 2291
Fax: 0208 534 2291
Website: www.poth-hille.co.uk
(Casting waxes)

POTTERYCRAFTS LTD (HEAD OFFICE)
Campbell Rd
Stoke-on-Trent
Staffs
ST4 4ET
Tel: 01782 745000
Fax: 01782 746000
Website: www.potterycrafts.co.uk
(Clay and clay tools)

TONY MAGUIRE PLASTERERS SUPPLIES
5 Barleycorn Way
Hornchurch
Essex
RM11 3JJ
Tel: 0845 65 86 677
Fax: 0845 65 85 329
Website: www.tomps.com
(Polyester resins and casting plasters)

TRAVIS PERKINS (HEAD OFFICE)
Tel: 0870 333 2111
Fax: 01604 588 294
Website: www.travisperkins.co.uk
(Casting plasters)

TRYLON LTD
Thrift St
Wollaston
Northants
NN29 7QJ
Tel: 01933 664960
Website: www.trylon.co.uk
(General mouldmaking and casting supplies)

INDEX